Studies in Systems, Decision and Control

Volume 641

The series "Studies in Systems, Decision and Control" (SSDC) covers both new developments and advances, as well as the state of the art, in the various areas of broadly perceived systems, decision making and control–quickly, up to date and with a high quality. The intent is to cover the theory, applications, and perspectives on the state of the art and future developments relevant to systems, decision making, control, complex processes and related areas, as embedded in the fields of engineering, computer science, physics, economics, social and life sciences, as well as the paradigms and methodologies behind them. The series contains monographs, textbooks, lecture notes and edited volumes in systems, decision making and control spanning the areas of Cyber-Physical Systems, Autonomous Systems, Sensor Networks, Control Systems, Energy Systems, Automotive Systems, Biological Systems, Vehicular Networking and Connected Vehicles, Aerospace Systems, Automation, Manufacturing, Smart Grids, Nonlinear Systems, Power Systems, Robotics, Social Systems, Economic Systems and other. Of particular value to both the contributors and the readership are the short publication timeframe and the worldwide distribution and exposure which enable both a wide and rapid dissemination of research output.

Indexed by SCOPUS, DBLP, WTI Frankfurt eG, zbMATH, SCImago.

All books published in the series are submitted for consideration in Web of Science.

Artur Zaporozhets · Inna Bilous · Ievgen Antypov

Buildings Energy Audit

Heat Loss, Air Exchange, Microclimate

Springer

Artur Zaporozhets
General Energy Institute
NAS of Ukraine
Kyiv, Ukraine

Inna Bilous
Igor Sikorsky Kyiv Polytechnic Institute
National Technical University of Ukraine
Kyiv, Ukraine

Ievgen Antypov
National University of Life
and Environmental Sciences of Ukraine
Kyiv, Ukraine

ISSN 2198-4182 ISSN 2198-4190 (electronic)
Studies in Systems, Decision and Control
ISBN 978-3-032-07199-6 ISBN 978-3-032-07200-9 (eBook)
https://doi.org/10.1007/978-3-032-07200-9

This Springer imprint is published by the registered company Springer Nature Switzerland AG
The registered company address is: Gewerbestrasse 11, 6330 Cham, Switzerland

Preface

Today, the issue of energy saving and energy security is of particular relevance both locally and globally. Today, the world is facing an acute problem of saving and rational use of energy resources in order not only to reduce their consumption but also to reduce pollutant emissions. For Ukraine, the issue of energy security has become particularly acute since the beginning of Russia's hybrid aggression. The destruction of energy infrastructure, the threat of disruption of energy supplies, and aggressive actions in the economic sphere are driving Ukraine to improve energy efficiency along the chain from production to consumption and to develop modern energy-saving technologies.

According to the European Parliamentary Research Service, households are the largest energy consumers in Europe, with final energy consumption accounting for 28% of the total energy consumption of all sectors. That is why researchers are focused on implementing resource-saving activities through the use of modern energy-efficient technologies in households and buildings, in particular, to reduce transmission and infiltration heat losses.

To determine the extent of heat transfer losses, it is common to measure the thermal resistance of buildings on site. These measurements are usually made by analysing the low-density heat flux.

To monitor the thermal insulation characteristics of building envelopes and materials used in repair work and establish their actual values, it is extremely important to develop modern monitoring tools and appropriate methods for measuring thermal resistance as the main indicator in assessing the energy efficiency of buildings.

Leading global organisations in the field of standardisation and certification, as well as research institutes and universities such as Munich University of Applied Sciences, DIN Deutsches Institut für Normung e.V., International Organisation for Standardisation, are constantly working on creating new methods and standards that would improve the accuracy of such measurements. The latest version of DSTU ISO 9869 "Thermal insulation, Building elements, Full-scale measurements of thermal resistance and heat transfer coefficient" has been harmonised in Ukraine.

It should be noted that measuring low-density heat flux is difficult for researchers due to the large fluctuation (up to 20%) of the heat flux during the experiment. Such difficulties are caused by various factors, one of the main ones being the high intensity of heat transfer. In addition, such measurements require the use of heat flux sensors with high sensitivity and special calibration for such a measurement range. It is well known that metrological aspects of measurements are very important and should be taken into account when determining and improving the parameters of the objects of study.

Today, the use of so-called "blower door testing" is growing in popularity around the world, which creates a pressure difference between the indoor and outdoor environment and stimulates the formation of air flows between the indoor and outdoor environment. For example, in the UK, France, Ireland and Denmark, airtightness testing of all new buildings or parts of them is required by regulations. In addition, the popularity of special energy programmes (e.g. Passivehaus, Minergie) that require or encourage airtightness testing of buildings is constantly growing. Currently, the leading air tightness testing systems manufacturers are BlowerDoor GmbH—Measuring Systems for Air Tightness.

An important component is the training of qualified personnel to carry out leakage tests. The number of specialists is growing rapidly in the UK, Belgium, Ireland, France and other European countries, as these countries require mandatory leak testing.

The air permeability of a building and its functional units is important both in terms of reducing energy consumption in building operation and in ensuring physiological and comfortable conditions for human occupancy. Reducing the air permeability of buildings through the use of very dense thermal insulation elements that are poorly permeable to air has serious consequences for indoor air quality due to increased concentrations of pollutants.

In this case, the lack of proper ventilation leads to a deterioration in sanitary and hygienic conditions in the room, so it is advisable to control the microclimate parameters additionally.

On 1 January 2023, the national standard DSTU EN ISO 9972:2022 Thermal performance of buildings. Determination of air permeability of buildings. Test pressure method (EN ISO 9972:2015, IDT; ISO 9972:2015, IDT). At the same time, an urgent task is to develop national practical recommendations (following the example of European countries) for conducting airtightness tests.

Increasing the airtightness of a building leads to a decrease in its air permeability due to the use of very dense thermal insulation materials that are poorly permeable to air. In this case, the lack of proper ventilation of the premises leads to a deterioration in sanitary and hygienic conditions, so it is advisable to conduct a comprehensive control of the building's air permeability and microclimate parameters.

The set of results on determining the main informative indicators of energy efficiency of buildings and conditions for conducting relevant research, improving methods of controlling air permeability and indoor climate will allow developing the structure of a national database for accumulating and processing experimental results, which will take into account the peculiarities of national and international standards for energy efficiency indicators of buildings, such as thermal resistance and air permeability, used in assessing the rational use of energy resources. This will make it impossible to ignore indoor comfort conditions during renovation and new construction.

The book consists of four chapters.

Chapter 1 analyses the current state and regulatory requirements for energy efficiency of buildings in Ukraine and the world, focusing on the impact of the energy crisis, particularly after the full-scale Russian invasion of Ukraine. A structural scheme of interaction of the main building subsystems in terms of their impact on the level of energy efficiency is proposed. The advantages of using BEM modelling to determine the energy demand, taking into account internal environmental indicators, are presented. Air exchange as an energy characteristic of a building is considered. The influence of air exchange on the quality of microclimatic characteristics of indoor air, in particular the level of CO_2 is analysed. An overview of pollutants in indoor air, their sources and the consequences of their impact on the body is given. Recommendations for the selection of metal oxide/electrochemical sensors for various gaseous substances (NO_2, CO, O_3) according to different metrological characteristics are formulated. Recommendations for improving the methodology of research on air permeability of buildings and indoor microclimate parameters, taking into account national and international standards and norms for testing, are proposed.

Chapter 2 considers the main informative indicators and models of energy efficiency of buildings and describes methods for assessing the condition of their external envelope and detecting thermal failures. The mathematical model of the thermal regime of a building as a single heat and power system is improved, taking into account its three subsystems: the model of heat and power impact of the external climate on the building, the model of heat and power impact of the thermal inertia of the building envelope, and the model of heat and power impact of the building's engineering systems. The necessity of conducting a thermal imaging inspection of buildings is substantiated, and the types of defects that can be detected during such an audit and the conditions under which the inspection will allow the detection of all defects are presented. Field studies of buildings for various purposes using the "Blower Door Test" and the thermal imaging method have been carried out, indicating the need to take into account the difference in pressure outside and inside the building.

Chapter 3 presents calculation methods for determining the natural actual air exchange for multi-apartment buildings based on the improved ASHRAE methodology. Quantitative estimates of the level of air exchange for a multi-storey building with different levels of air permeability of windows and the impact of dynamic changes in climatic conditions on the level of natural air exchange are carried out. The hourly changes in the natural air exchange rate for rooms of different floors and orientations of the studied representative building are calculated. An improved hourly dynamic grid model for determining the energy demand of representative premises based on the EN ISO 13790:2013 and EN ISO 13786:2007 standards is proposed.

Chapter 4 presents a study of natural air exchange based on experimental data on changes in the level of CO_2 concentration in the premises of educational institutions and a residential building. The results of dynamic modelling in the Design

Builder software are presented. The parameters affecting the natural component of air exchange are analysed, and spatial dependence curves are established. Attention is focused on considering seasonal variations when calculating air exchange. It is shown that in summer, natural air exchange mainly occurs under the influence of the wind effect, and during the heating period, air exchange mainly occurs due to a combination of the stack and wind effects. The level of energy consumption of the studied premises is determined, provided that the normative air exchange established by national and international standards is observed and the actual air exchange rate determined experimentally is considered.

The book is intended for researchers, engineers, teachers, postgraduates and students of higher educational institutions dealing with issues of increasing the energy efficiency of buildings and structures, energy management, and energy auditing.

Kyiv, Ukraine
December 2024

Artur Zaporozhets
Inna Bilous
Ievgen Antypov

Competing Interests The authors have no competing interests to declare that are relevant to the content of this manuscript.

Contents

Chapter 1
Current State and Analysis of Regulatory Requirements for Energy Efficiency of Buildings

1.1 Current State of Energy Efficiency in Buildings in Ukraine and the World

The issue of energy efficiency in Ukraine has become particularly acute since the full-scale invasion and destruction of energy-generating facilities. That is why energy saving is now coming to the fore both in the short and long term. In particular, improving the energy efficiency of the construction sector as a significant energy consumer. According to the GMK Center analytical resource, electricity, gas and air conditioning together account for 52% of Ukraine's CO_2 (carbon dioxide) emissions by industry. The construction sector is one of the largest consumers of these resources.

According to the general characteristics of the housing stock, as of 1 January 2020, there were more than 8 million residential buildings in Ukraine with a total living space of 892.15 million square metres^2 [1].

The efficiency of fuel and energy use in Ukraine's housing stock is unsatisfactory. According to the Energy Balance of Ukraine in 2020, the residential sector was the second largest energy consumer, as shown in Fig. 1.1. According to the statistics, the share of the residential sector of 13.601 million tonnes of oil equivalent is 28.5% of the total final energy consumption in the country for the period.

According to the State Standard, in the structure of energy consumption by households by purpose in 2020, residential heating accounted for 53% of all energy sources. Households accounted for 28.5% and the service sector for another 10.2% of total final energy consumption in 2020. According to the Ministry of Regional Development, the residential real estate sector has significant economic potential, which can reach 9 million tonnes of oil equivalent annually with reduced energy consumption by residential buildings.

According to the State Statistics Service of Ukraine in its Statistical Digest "Socio-Demographic Characteristics of Ukrainian Households in 2019", as of 2019, almost half of households (44.5%) had not carried out major repairs to their housing (in urban areas, the figure is 53.3%). In addition, in 3.1% of households, the last major

A. Zaporozhets et al., *Buildings Energy Audit*, Studies in Systems, Decision and Control 641, https://doi.org/10.1007/978-3-032-07200-9_1

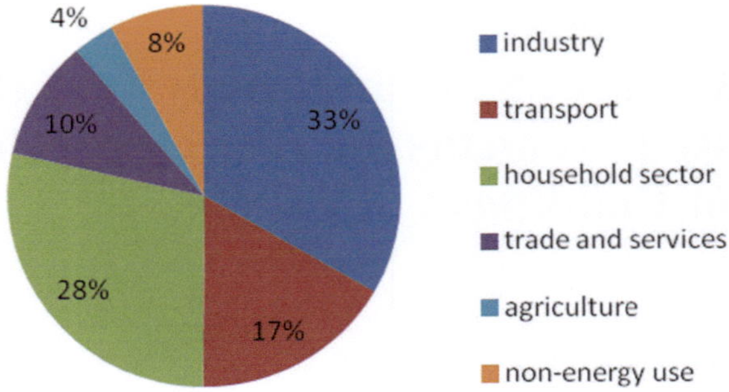

Fig. 1.1 Energy balance of Ukraine for 2020

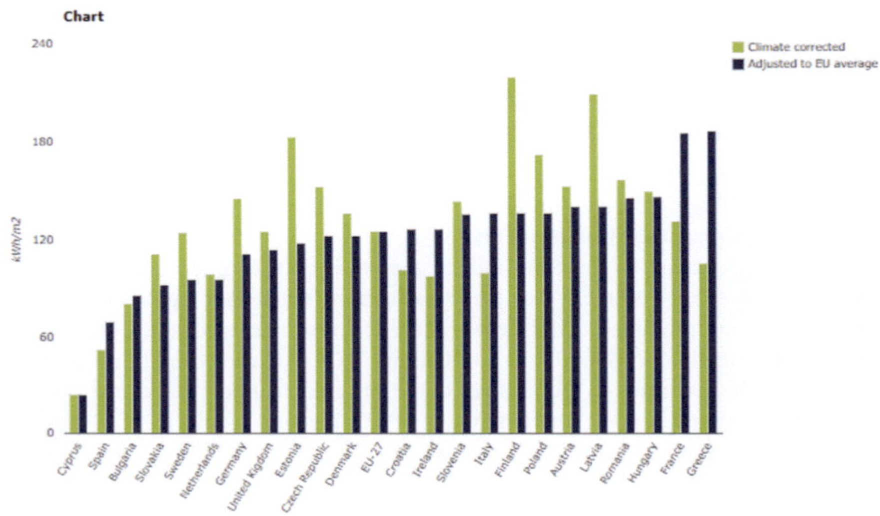

Fig. 1.2 Household energy consumption for heating in the EU countries [4]

repair took place before 1990. It is important to note that according to 2017 statistics, 87.2% of households in Ukraine as a whole live in housing built before 1991 (86.6% in urban areas). This suggests that the absence of major repairs in almost half of Ukrainian households may indicate the insufficiency and incompleteness of measures needed to improve their housing, including energy efficiency. The insufficient thermal efficiency of residential buildings, which arises from the lack of major repairs, leads to significant heat losses (up to 50%). According to the results of energy surveys, heat losses in residential buildings built in the past years through walls are 42%, through windows—16%, through the roof—7%, through the basement—5%, and during air exchange—30% [2]. The actual efficiency of heat energy use in Ukrainian buildings is 3–5 times lower than in Western countries [3].

Experts have found that in Ukraine, energy consumption for heating and hot water per unit area is significantly higher than in European countries with a similar climate. For example, the average energy consumption for heating in EU countries is 130 kWh/m^2, while in Ukraine this rate reaches 264 kWh/m^2 [3].

The European Union recognises that improving energy efficiency is one of the key conditions for achieving sustainable development. In order to improve the energy efficiency of construction projects in the EU, the requirements for final energy consumption are constantly increasing. Starting from 2019, it is prohibited to build residential buildings with consumption exceeding 60 kWh/m^2 for similar climatic conditions to Ukraine [5].

Since 2020, Europe has been undergoing a large-scale transition to the construction of buildings according to NZEB (Nearly zero-energy buildings) standards, i.e. buildings with almost zero energy consumption. The energy consumption of such buildings should not exceed 15 kWh/m^2.

In 2020, the Cabinet of Ministers of Ukraine developed and approved the Concept for the Implementation of State Policy in the Field of Energy Efficiency of Buildings.

According to the State Agency on Energy Efficiency and Energy Saving, comprehensive thermal modernisation of residential buildings in Ukraine will reduce energy consumption by half compared to the current level [5].

State control over the level of energy use in the construction industry is exercised by assessing the energy performance of buildings. Thus, the assessment of energy consumption during design, construction and operation has become the main determinant of their energy efficiency. Estimated annual energy consumption is used for energy certification and preparation of building passports, selection of investment projects to improve energy efficiency, planning future energy costs and demonstration of compliance with specific energy performance requirements and for demonstrating compliance with specific energy performance requirements [6].

The European Union is working to significantly reduce energy consumption, contributing to the global goal of becoming climate neutral by 2050.

Ukraine has defined a sustainable development strategy until 2030, which envisages a 15% reduction in energy use in buildings through thermal modernisation. Sustainable building professionals also support the modernisation of outdated facilities, as renovation is proving to be a more cost-effective process than building new green buildings to replace them. This process includes thermal insulation of walls and roofs, replacement of windows, installation of solar panels, ventilation systems with heat recovery, and other measures [7].

According to the Law of Ukraine “On Energy Efficiency”, newly constructed or renovated facilities must have an energy efficiency class of at least “C” (there are seven classes, with “A” being the highest and “G” the lowest). In general, the lowest energy efficiency class G among all certified buildings is residential buildings (47.2%), educational institutions (59.7%) and healthcare facilities (46.2%) [8].

In January 2024, the State Fund for Decarbonisation and Energy Efficient Transformation was launched in Ukraine. This new legal source of funding will provide additional funds for the implementation of energy efficiency and carbon reduction programmes in the country.

1.2 Interaction of Building Elements as Components of the Energy System

One of the most important requirements for modern buildings is the level of their energy efficiency [9], i.e., the minimum energy consumption to maintain thermal comfort in the premises, inextricably linked to the automatic regulation of heating systems [10]. This is primarily due to the understanding and practical confirmation of the essential role of automatic control in reducing the heat consumption of buildings and bringing it to the normalised value [11]. In order to meet the necessary requirements for ensuring the internal microclimate parameters of the premises [12] (Table D.4), it is necessary to properly design the thermal protection properties of external enclosures, calculate the thermal capacity of the heating system, the starting point of which will be the estimated external conditions [13]. The design parameters of the microclimate of premises depend on their functional purpose and sanitary and hygienic requirements. For most residential and public buildings, these conditions are approximately the same. At the same time, the determining factor in improving the energy efficiency of a building is the study of the mutual influence of these parameters, taking into account the provisions of [14, 15].

The results of the study of the influence of thermal inertia of a room are presented in [16], where the consideration of the internal heat capacity of a building, on the example of the educational and administrative building of the National University of Life and Environmental Sciences of Ukraine (NULES of Ukraine), allows avoiding energy overruns of about 10–12%. In this direction, we have studied the change in the comfort conditions of premises from the inertia of the external building envelope when the heating system operates in energy-saving modes [17], as well as the dependence of the comfort conditions of public buildings with different degrees of thermal protection of external building envelopes when the heating system operates in the standby mode [18], and its impact on the dynamics of energy consumption.

The thermal comfort of a building after its thermal modernisation using renewable energy has been studied [19, 20] and taking into account the provisions of relevant standards, norms and methodologies [21–23].

The scheme of interaction of building elements as an energy system is shown in Fig. 1.3. The main influence on the formation of thermal comfort and, accordingly, the level of energy efficiency of a building is exerted by its thermal insulation envelope, the properties of which determine the choice of parameters of the heating subsystem.

The adequacy of the assessment of the energy efficiency of buildings and the detailing of energy performance indicators largely depend on the use of mathematical models of the building, the complexity of which is determined by the tasks to be solved. Trends in the development of mathematical models for determining the energy state of a building are aimed at reducing time intervals and at determining the localisation of system characteristics and influencing factors [25]. Modern computer methods and tools for building energy modelling based on BEM models (Building Energy Modelling) are a universal, multi-purpose tool that allows for real-time energy consumption management. BEM is used in the in-depth analysis of a building's

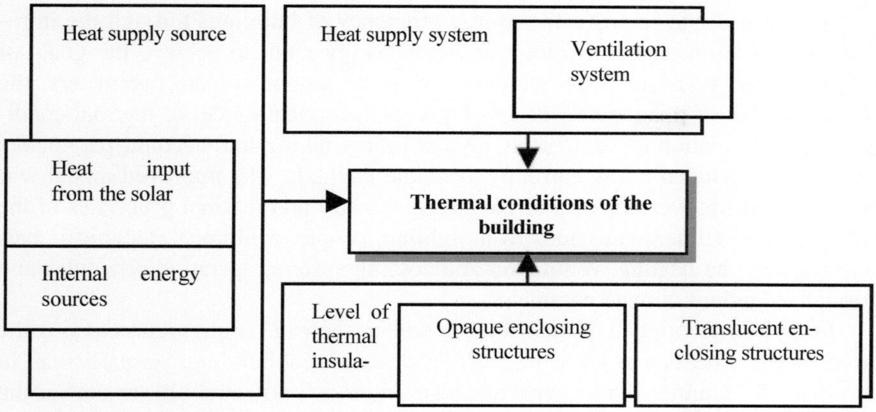

Fig. 1.3 Interaction of the main building subsystems in terms of their impact on energy efficiency [24]

energy efficiency and the development of strategies (energy efficiency policies). The main aspect of BEM modelling is the determination of energy demand, taking into account internal environmental indicators (thermal and visual comfort, indoor air quality, etc.) [26]. Among the most powerful software products implemented on the principles of BEM modelling are BLAST, DOE-2, ESP-r, HVACSIM+, TRNSYS, EnergyPlus and others [27]. BEM modelling allows us to consider a building as an energy system, which is illogical to European standards, which are being actively implemented in Ukraine. Here, a building, as a complex energy system, is considered as a combination of the external climate, engineering systems, envelope, and human, as an indicator of comfort conditions and energy processes in the building's premises.

One of the most popular approaches to assessing the energy performance of buildings is to use EnergyPlus software together with the DesignBuilder user interface. The mathematical formulation is based on a system of equations for heat transfer between indoor air and the external environment through transmission heat transfer and radiation heat transfer of enclosures, taking into account non-stationary energy storage processes in building structural elements [28]. However, the model does not consider the influence of the external climate parameters that are crucial for winter, such as the outside air temperature $t_{з.n}$, and wind speed υ. In some calculations, the relative humidity φ, and enthalpy i, of the outside air, as well as solar radiation, wind direction, should be taken into account in addition to them, adjusted for the height of the building, but this is not possible in the existing building model.

To assess the level of thermal comfort, the method presented in the ISO 7730 standard is widely used, based on the heat balance equations for the human body [29]. However, it does not specify the mechanism of thermoregulation, which is essential when calculating human heat transfer.

Thus, in order to improve the energy efficiency of buildings through the introduction of information technology for their energy audit to achieve the goals of reducing energy consumption while ensuring the indoor climate parameters, the second section of this report will develop a mathematical model of thermal conditions and calculation of heat losses by a building, taking into account parameters that are not included in the currently available methods. The improved model will take into account weather data of the area, geometry and thermal properties of the building, internal heat emissions from lighting, people, equipment, technical characteristics of the heating, ventilation and cooling system, operation schedules and normalised indoor climate parameters.

Taking into account all of the above, as well as the wide configuration among the most common heating systems in public and residential buildings, in particular, in buildings of Ukrainian higher education institutions, it is also advisable to analyse the impact of the building's engineering systems on its thermal regime and heat energy savings, provided that the degree of their automation is brought to the minimum permissible level "C", which is defined by the provisions of [9–12, 14, 15].

1.3 Air Exchange as an Energy Characteristic of Buildings

According to international norms and standards, which are being actively integrated into Ukrainian ones, buildings are considered integrated energy systems. These systems consist of building envelopes, energy sources, engineering systems, and also take into account the impact of the external climate. All these elements interact to provide comfortable conditions for people living or working in the building, while using energy resources efficiently [30].

When studying the energy efficiency of buildings located in temperate and cold climates, the main focus is on the heat consumption of buildings. This is because heat consumption is a key element of energy costs. In addition, it is the buildings sector that has the greatest potential for energy savings [30].

The thermal and energy performance of buildings is determined by numerous external and internal factors.

An effective analysis of energy efficiency measures should take into account the dynamic behaviour of the building—the relationship between the internal temperature regime and the factors that influence it. In Ukraine, the use of dynamic models to determine the energy performance of buildings using hourly climate data is becoming increasingly relevant [6, 30].

Energy consumption in a building is influenced by six main factors: external climate conditions, building envelope design, equipment and energy sources, systems, operations and maintenance, occupant activities and behaviour, and indoor environmental conditions.

The standards for calculating the thermal capacity of the heating system of multistorey residential buildings are determined by the current state building codes—DBN B.2.2-15:2019 "Residential buildings. Main provisions" [5].

In many European countries, the energy efficiency of buildings is addressed at the design stage. Buildings are designed with the latest technologies and standards in mind to minimise energy consumption and reduce environmental impact. However, in Ukraine, most existing buildings were built several decades ago and do not meet modern energy efficiency standards. Therefore, the analysis of energy efficiency indicators is carried out at the stage of operation of existing buildings, which has its own peculiarities, as it requires taking into account many factors.

The results of energy analysis of buildings of various types in Ukraine and abroad indicate that after the implementation of insulation measures, the share of costs associated with ventilation air is the highest [31], provided that the requirements for air exchange are met. The ventilation component of heat loads in residential and public buildings is 30–50% [32]. In the USA, researchers conclude that the infiltration component of the cost reaches 33% [32].

The main areas for improving the energy efficiency of buildings are the use of effective thermal insulation during construction and reconstruction, as well as the use of windows with high thermal protection properties and low air permeability. These measures lead to an increase in the airtightness of residential and public buildings, which makes it difficult for fresh air to enter the premises. To solve this problem, ventilation is often carried out or mechanical ventilation is installed, which reduces the effectiveness of the implemented energy saving measures. Thus, there is a conflict between energy saving and comfort, which requires optimal air exchange. To resolve this conflict of interest, it is necessary to consider the use of modern ventilation systems that provide efficient air exchange without significantly affecting the energy efficiency of the building.

Apartment buildings pose unique challenges to indoor air quality (IAQ), as pollutants can move from one room to another and residents have limited ability to make changes to the building structure itself.

For example, the United States has created the Standard Work Specifications (SWS), a free online tool and industry guide that defines minimum acceptable outcomes for home energy retrofits. The SWS and the U.S. Environmental Protection Agency's (EPA) Healthy Indoor Environments for Residential Energy Retrofits protocols were developed jointly and are intended for voluntary adoption by federally funded housing programmes, private sector contractors, and other organisations and entities involved in residential energy retrofits. The SWS reflects a whole-building approach to implementing energy efficiency measures, which may include ventilation, insulation, and air sealing [7].

In the majority of existing buildings in Ukraine, mechanical ventilation is either not provided for in the design or does not work properly. Exhaust air is removed directly from the areas where it is most contaminated (kitchens and bathrooms) through a natural exhaust duct system. Fresh air enters the premises by infiltration through various gaps and openings in the building envelope (mainly through window structures) or as a result of airing and is heated by the heating system. In this way, air is exchanged in all areas of the building. However, after replacing old windows with new, sealed metal-plastic ones, the air regime in the apartments is disturbed. This is due to the fact that in the absence of a mechanical ventilation system, the infiltration

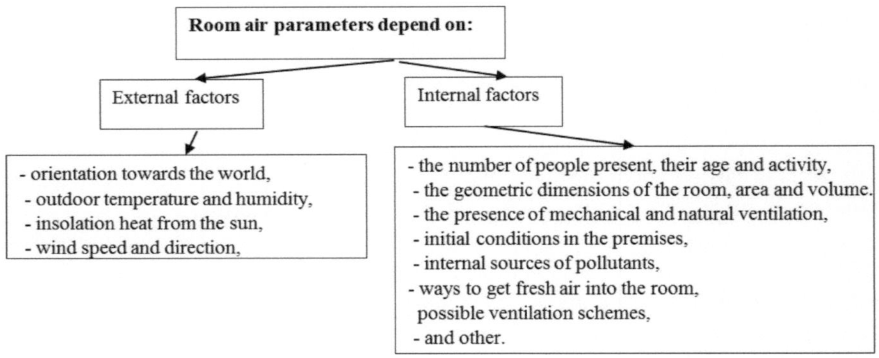

Fig. 1.4 Indoor and outdoor air parameters of the room

of outside air becomes the only source of fresh air. This can lead to the need to install additional ventilation systems to ensure optimal air exchange.

Infiltration is a form of natural ventilation where air is exchanged in a room through channels that are not specially organised. It is caused by the difference between the atmospheric pressure and the pressure inside the room, as well as by wind penetrating through gaps in structures such as windows, doors, ceilings, and when vents, transoms and doors are opened. ASHRAE (American Society of Heating, Refrigerating and Air-Conditioning Engineers) defines air infiltration as the uncontrolled leakage of outside air into a building caused by the presence of a pressure differential between different elements of the building envelope [8].

Natural ventilation is one of the most difficult indicators to determine, as it depends on a large number of factors. All factors, both internal and external, affect the efficiency of natural ventilation and require a detailed analysis to optimise the energy efficiency of a building (Fig. 1.4).

Determining the air exchange rate (AER), taking into account the influencing factors for the operating conditions, is a complex task. Typically, professionals are guided by the values specified in standards. However, to date, most standards do not define a specific method for establishing the minimum level of air exchange, but instead offer different strategies for its calculation. As a result, experts, based on national standards, draw their own conclusions about the method of determining the normative air exchange in the premises. Experts use both empirical and experimental methods to determine the air exchange rate. The most common empirical calculation methods are based on ASHRAE and BLAST standards (analysis of building loads and system thermodynamics) and determine air exchange through pressure differences [33, 34]. Experimental measurements are relevant because of the possibility of obtaining actual data in real time [30, 34–36]. Empirical methods have been used in the development of software for calculating the energy performance of buildings.

Also, the study of the infiltration component is important not only in terms of its impact on energy consumption, but also as an important component of indoor air quality.

Therefore, the study of the air exchange rate in buildings is relevant both in the operation and thermal modernisation of buildings, as the use of up-to-date data on air exchange parameters in buildings will allow the development of effective strategies to reduce energy consumption in buildings, taking into account the dynamics of the operational characteristics of buildings and ensuring the requirements for indoor air quality.

1.4 Impact of Air Exchange on Indoor Air Quality

Heating, ventilation and air conditioning (HVAC) systems are designed to maintain thermal comfort and microclimate quality in buildings. The HVAC system often accounts for a large percentage of the total cost of energy expenditures in a building. In the United States, this figure is about 40% [37], and in the post-Soviet countries, this figure is over 40%. Given the global trends aimed at reducing energy consumption and greenhouse gas emissions, software is used to model energy processes in buildings, which allows for consideration of various design and operation options. In the countries of the former Soviet Union, including Ukraine, a significant number of buildings belong to the mass construction of the 1980s, which is characterised by a low level of energy efficiency.

The energy performance of buildings is influenced by the thermal and physical properties of the envelope, geometry, heat gain from the sun and additional internal heat, heating, air exchange and operating features. Based on studies [38, 39], the most influential parameter is the air exchange rate.

In most cases, energy efficiency improvements start with building thermal renovation measures [40, 41]. Reducing energy consumption and ensuring comfort conditions [42] is an important synergistic effect when implementing energy saving measures. In accordance with the Ukrainian standard [43], the recommendations on the priority of energy saving measures, the main focus is on measures for thermal sanitation of the building envelope. After the implementation of thermal sanitation measures (the first stage), the ventilation component of heat loads in residential and commercial buildings can be from 25 to 50% [44, 45]. Reducing this component is the second stage of energy saving measures in Ukraine [46].

The European Committee for Standardisation (CEN) developed the European Standards (EN) for building ventilation in 2007. Recommendations for minimum outdoor air flow rates per person in public and residential buildings can be found in the European standards EN 16,798-1:2019 [47] and EN 16,798-3:2019 [48].

In addition, the United States has adopted ASHRAE 62.2-2022 (Ventilation and Acceptable Air Quality in Residential Buildings) [49] and ASHRAE 62.1-2022 (Ventilation and Acceptable Indoor Air Quality) [50]. ASHRAE 62.1-2022 [50] specifies minimum ventilation rates and other measures designed to provide indoor air quality that is acceptable to occupants to minimise adverse health effects. The standard provides procedures and methods for meeting minimum ventilation and indoor air quality requirements for engineers, designers, owners, and jurisdictional authorities. ASHRAE Standard 62.2-2022 [49] describes the minimum requirements for achieving acceptable IAQ through residential ventilation, local mechanical exhaust, and source control for residential occupancies with temporary occupants. ASHRAE is accredited by the American National Standards Institute (ANSI) and follows ANSI's requirements for due process and standards development. All standards above provide default values that can be used when national legislation is not available.

The above new ventilation intensity standards take into account both the presence of people and internal sources associated with the building. A building is characterised as low pollution if the building materials used for interior surfaces meet national or international criteria for building materials. The required ventilation rates are suggested based on health and comfort criteria.

Recommended input values are provided for each of the different categories. Below is a brief description of the categories.

- Category I: high level of expectations and is recommended for spaces occupied by very sensitive and fragile people with special requirements, such as the disabled, sick, very young children and the elderly.
- Category II: Normal level of expectations and should be used for new buildings and renovations.
- Category III: acceptable, moderate level of expectations and can be used for existing buildings.
- Category IV: Values outside the criteria for the above categories. This category should only be accepted for a limited period of the year.

Regulations for residential and public space ventilation vary from country to country. According to the European regulations Directive 2010/31/EU and Directive 2012/27/EU, updated in 2018 and 2019 respectively, European Union countries must use national energy efficiency requirements that allow for international comparison. National standards in EU countries do not specify a specific approach to ventilation in residential and public buildings. In most countries, both natural ventilation and mechanical or hybrid ventilation systems can be used. Air flow is defined in l/s per m^2, l/s per person or through the air exchange rate per $hour^{-1}$ [2].

In addition to international standards, there are ventilation requirements in the national building codes of many EU countries. In Ukraine, EN 15,251:2011 and EN 16,798–3:2019 (replacing EN 13,779:2011) have been introduced as national standards [3, 47].

The EN 15,251:2011 standard states that there is no general standard indicator for determining indoor air quality. In this case, indoor air quality is defined as the required level of ventilation or CO_2 concentration. It is generally accepted that indoor air quality is influenced by pollution from people and their activities, from buildings and their equipment, and from the HVAC system itself. The latter two sources are commonly referred to as building components, but there is no general agreement on how to add the different sources of pollution [51]. The current standard provides several methods for determining the required ventilation rate.

In accordance with the method based on the ventilation rate per person or per m^2 of total floor area, Tables 1.1 and 1.2 provide recommended ventilation rates for non-residential and residential buildings, respectively.

According to the ASHRAE 62.2-2022 standard, you can use general table values of air exchange rates for the entire apartment (Table 1.3), which are selected depending on the size of the dwelling and the number of bedrooms.

According to the table values of ASHRAE 62.2-2022, for a one-room apartment with a total area of 32.8 m^2, the total ventilation rate for housing will be 14 l/s per m^2. With an internal ceiling height of 2.5 m in all rooms, the equivalent air exchange rate will be 0.61 h^{-1}.

In the process of energy certification of buildings, energy auditors assume that the air exchange rate in residential buildings with natural ventilation is 0.6 h^{-1} in accordance with the (state standard of Ukraine) DSTU 9091:2022 "Energy Efficiency of Buildings". However, in the calculations of the energy certificate of buildings, this value is taken as infiltration. DSTU 9190:2022 [52] refers to DBN B.2.5-67:2013 [53], which in turn refers to DSTU B EN 15,251:2011.

Ukrainian state building codes for residential buildings DBN V.2.2-15:2019 [5] are applied for design, reconstruction and overhaul conditions, and contain ventilation

Table 1.1 Examples of recommended ventilation rates for non-residential buildings in three pollution categories from the building alone according to EN 15,251:2011 [51]

Category	Air flow per person, $\frac{l/s}{pers}$	Airborne emissions of building pollution, $\frac{l/s}{m^2}$		
		Very low level of pollution in the building	Low level of pollution in the building	High level of pollution in the building
I	10	0.5	1	2
II	7	0.35	0.7	1.4
III	4	0.2	0.4	0.8

Table 1.2 Examples of recommended ventilation rates for residential buildings EN 15,251:2011 [51]

Category	Air exchange rate[a]	Multiplicity	Living room and exhaust air flow, l/s, bedrooms, mainly outside air flow				
	$\frac{l/s}{m^2}$	h^{-1}	$\frac{l/s}{pers^b}$	$\frac{l/s}{m^2}$	Kitchen[a]	Bathrooms[b]	Toilet facilities
I	0.49	0.7	10	1.4	28	20	14
II	0.42	0.6	7	1.0	20	15	10
III	0.35	0.5	4	0.6	14	10	7

[a] The air exchange rate is expressed in l/s, m^2 and the air exchange rate correspond to each other at a ceiling height of 2.5 m

[b] The number of occupants can be calculated from the number of bedrooms. If necessary, national assumptions should be used; for energy and indoor air quality calculations, these may vary

Table 1.3 Requirements for ventilation air for residential buildings in accordance with ASHRAE 62.2-2022, l/s [50]

Total area, m^2	Number of bedrooms				
	1	2	3	4	5
<47	14	18	21	25	28
47–93	21	24	28	31	35
94–139	28	31	35	38	42
140–186	35	38	42	45	49
187–232	42	45	49	52	56
233–279	49	52	56	59	63
280–325	56	59	63	66	70
326–372	63	66	70	73	77
373–418	70	73	77	80	84
419–465	77	80	84	87	91

requirements for calculating heat loss in the form of minimum values of air exchange rates and air flow rates for premises for various purposes [54].

The numerical values of the requirements for air exchange in residential buildings according to national and international standards are given in Table 1.4.

The current national standard DBN B.2.2-3:2018 sets out the requirements for air exchange in premises for various purposes in educational institutions. According to the standard [55], classrooms, classrooms and laboratories of general secondary education institutions and classrooms of the first to fourth grades require air exchange of 16 m^3/h per person. For auditoriums, classrooms and vocational schools and universities (higher education institutions), training workshops with areas for theoretical knowledge, reading rooms, course design rooms, painting studios, and a number of other educational premises, the air exchange requirements are 20 m^3/h per 1 person.

Table 1.4 Requirements for air exchange in residential buildings according to national and international standards

Premises	DBN V.2.2-15:2019	DSTU 9190:2022, DBN B.2.5-67:2013 with reference to DSTU B EN 15,251:2011	ASHRAE 62.2-2022
Common room, bedroom, children's room, study	0.50	0.60	0.61
Kitchen, kitchen-dining room	1.50		
Corridor	0.50		

Table 1.5 shows the supply air and air exchange rates for classrooms from some international and national standards and national standards of some European countries. These values were derived from a typical classroom situation with 25 students, an area of 50 m^2 and an internal height of 3 m, which corresponds to a volume of 150 m^3 [56].

Analysing the data presented in Table 1.4, it can be concluded that Ukraine has the lowest requirements for air exchange in classrooms compared to a number of other countries.

The absence of a clear approach in most standards for establishing normative air exchange forces designers to choose their own strategy for its calculation based on the recommendations of these standards. When performing energy calculations, the air exchange requirements defined in national standards are usually used.

The list of standards and norms in force in Ukraine that specify the requirements and norms for air exchange in residential and public buildings as of 2024:

- DSTU B EN 15,251:2011 "Estimated parameters of indoor microclimate for the design and assessment of energy performance of buildings in terms of air quality, thermal comfort, lighting and acoustics";
- DSTU B EN 15,243:2015 "Ventilation of buildings. Calculation of room temperature and methods for determining loads and energy requirements for buildings with air conditioning systems";
- DSTU 9190:2022 "Energy efficiency of buildings. Method for calculating energy consumption for heating, cooling, ventilation, lighting and hot water supply";
- EN 15,242 "Ventilation of buildings. Methods for calculating air flow rates for building ventilation with consideration of infiltration";
- EN 15,241 "Ventilation of buildings. Methods for calculating energy losses from ventilation and air infiltration in public buildings";

Table 1.5 Requirements for air exchange in educational premises [56]

Country (standard or norms)	Air exchange	
	m/h^3	h^{-1}
[ASHRAE 62.1 (2022)]	558	3.7
[EN 15,251 (2007)] (a)	756	5.0
Ukraine [DBN V.2.2–3:2018]	400 (500) (b)	2.7 (3.3) (b)
Portugal [RECS (2013)]	600	4.0
United Kingdom [Building Bulletin 101 (2006)]	450 (d)	3.0
Germany [DIN1946-2 (2005)]	500	3.3
Finland [National Building Code—Part D2 (2010)]	540	3.6
France [Règlement Sanitaire Departmental Type (2004)]	375–450	2.5–3.0

(a) Values for class II (normal expectation level)
(b) Requirements for classrooms of higher education institutions
(d) Daily average; the upper recommended limit is 720 m^3/h. For naturally ventilated classes, the minimum supply air volume requirement is 270 m^3/h

- DSTU EN 16,798–3:2019 "Energy efficiency of buildings. Ventilation of buildings. Part 3. Ventilation in non-residential buildings. Operational requirements for ventilation and air conditioning systems in the premises";
- DBN B.2.2–3:2018 "Buildings and structures. Educational institutions";
- DSTU B CEN/TR 14,788:2015 "Ventilation of buildings. Design and characterisation of ventilation systems for residential buildings";
- DBN B.2.2–15:2019 "Buildings and structures. Residential buildings. Main provisions".

In the EU countries, as well as in Ukraine, standard values for the air exchange rate for the building as a whole are often used in energy audits of buildings. However, national standards also provide for the possibility of determining the air exchange rate in individual rooms of a building based on special tests.

There are a number of international standards that can be used to guide the testing of air exchange in buildings, some of which have been implemented as national standards:

- DSTU EN ISO 9972:2022 "Thermal performance of buildings. Determination of air permeability of buildings. Test pressure method" (replacing DSTU B V.2.2-19:2007);
- DSTU EN ISO 20485:2019 "Non-destructive testing. Tightness control. Indicator gas method";
- ISO 12569:2017 "Thermal performance of buildings and materials. Determination of specific air flow rate in buildings. Method of dilution of indicator gas";
- ASTM E741-23 Standard Test Method For Determining Air Change In A Single Zone By Means Of A Tracer Gas Dilution. Selection of methods and models for determining air exchange.

At the national and regional levels, one of the most targeted segments is public buildings, with special attention paid to educational institutions. Recently, much attention has been paid to the ventilation component (to determine the air exchange rate), which can account for 30–50% of total energy consumption [57, 58]. To ensure proper working conditions in terms of ventilation, the standard [53] regulates the standard air exchange rate. In a building, ventilation is provided in two ways: natural and mechanical. In most old Ukrainian educational buildings, artificial ventilation is not provided or does not work. Thus, due to the deterioration of the main building stock, ventilation comes through air ducts, ventilation and leaks in windows, doors, etc. The peculiarities of implementing energy-saving measures in Ukraine are in line with the general trends in Central and Western Europe, where, after the implementation of measures to improve the thermal properties of building envelopes, the concentration of CO_2 in naturally ventilated premises has increased significantly [59]. Reducing this component is the second step in the implementation of energy saving measures in Ukraine. Therefore, in the context of the widespread implementation of programmes aimed at improving the thermal protection properties of the building envelope, more attention should be paid to the factors affecting natural ventilation (free convection).

Air exchange is affected by a large number of parameters that can be divided into internal and external [60].

Accurate accounting of the infiltration component is the key to the reliability and quality of calculations for determining the feasibility of implementing energy-saving measures. The natural air exchange calculation is based on using physical and empirical models [61].

The intensive use of educational buildings leads to a significant increase in the concentration of CO_2 in the premises, which requires additional research. Carbon dioxide (CO_2) is one of the indicators of human biological excretions. A classification is generally accepted and used in standards for human-occupied spaces where the main pollution is caused by human metabolism [62–64]. At the same time, determining carbon dioxide content with modern portable devices is a simple and convenient method for determining suffocation and, based on these measurements, ventilating school classrooms and auditoriums.

The permissible values for the content of CO_2 in classrooms in comparison with the requirements of the standards of other countries and Ukraine are given in Table 1.6 [65, 66], which differ slightly from each other.

The determination of the actual air exchange and CO_2 concentration in a room has received considerable attention from scientists [67–70]. There are various methods for measuring AER, such as SF6 tests [69], but they are usually quite complex and expensive. A number of studies have been conducted based on a full-scale experiment to determine the natural air exchange rate for buildings in South China and Europe [71, 72], which requires special equipment and provides an average view of air exchange, taking into account the variability of internal and external parameters that affect air exchange over time. A convenient method for monitoring air exchange rates using continuous CO sensors$_2$ has been developed in both laboratory experiments and field studies in classrooms in China [57]. Similar approaches were used in articles [58, 73]. The results obtained by such methods should be supplemented and analysed using mathematical modelling. Similar approaches to determining natural air exchange were used on livestock farms in Germany [68], with the peculiarity that the buildings where the animals were kept were single-storey, long-shaped with large doorways [74].

Article [67] notes that residential buildings in Central and Eastern Europe do not meet modern energy efficiency requirements. In addition, the quality of indoor air is not given due attention. The authors of [67] studied the concentration of CO_2 in a residential building in Slovakia before and after renovation. It was found that after the renovation of buildings, air exchange is significantly reduced and the level of CO_2 increases. Therefore, the implementation of measures to improve the thermal protection properties of the building envelope should be combined with measures for mechanical ventilation [74].

For schools in Central Spain with natural air exchange [70], which occurs by opening windows, a study of CO_2 concentrations in classrooms for the heated and air-conditioned period was conducted, which established recommendations for airing classrooms, which can vary from 5 to 20 min depending on a number of internal and

Table 1.6 Comparison of standards to the content requirements of CO_2 in classrooms

	Country	Rate	CO level$_2$
1	Finland	Standard of the Ministry of Health and Social Development, 2003	Air quality:
			High—700 ppm
			Average—900 ppm
			Satisfactory—1200 ppm
2	USA	Recommendations on air quality in schools of the US Department of Health and Human Services	Threshold level
			1000 ppm
3	USA	ASHRAE 62-1989 "Ventilation for normal air quality"	1000 ppm
4	USA	Recommendation of the Occupational Safety and Health Administration (OSHA), 1994	800 ppm
5	USA	Recommendation of the American Association of Industrial Hygienists (ACGIH), 1998	600 ppm
6	Russia	GOST 30,494-2011	Optimal values
			500–800 ppm
			Limit of acceptable
			1400 ppm
7	United Kingdom	"Ventilation in school buildings. Guidelines for Standards and Design", 2006	1500 ppm—the maximum allowable level for the school day from 9.00 to 15.30
8	Netherlands	Hygiene standards "Overview of air quality standards for kindergartens in the Netherlands"	1000 ppm—hygienic standard for kindergartens; 1200 ppm—hygienic standard for schools
9	Estonia	Regulations of the Ministry of Social Affairs	1000 ppm is the hygienic standard for schools
10	Germany	DIN EN 15,251, input parameters for indoor climate for design and assessment of the energy performance of buildings—indoor air quality. Temperature, light and acoustics. 2007 p.	Absolute concentration for energy efficiency calculations 750...1200 ppm
11	Ukraine	DBN B.2.2-3:2018 with reference to DSTU B EN 15,251:2011	750...1200 ppm

external factors. Similar problems with high levels of CO_2 are inherent in Ukrainian schools [74].

Physical models for determining natural air exchange are quite complex and require a large number of input parameters, which makes the calculation of buildings with a large number of zones significantly more complicated. Empirical methods are the most common. Empirical approaches for estimating the natural component of air exchange have the advantage of being easy to use for multi-zone building models. These approaches use algebraic equations that relate building features such as elevation, orientation, air permeability of structures, and weather conditions. One of the first approaches was developed by Shaw and Tamura [60], which was based on a single equation that combined the stack effect and wind effect to calculate the natural air exchange rate.

Empirical approaches are reflected in software systems for calculating the energy performance of buildings. The most widely used software packages for energy modelling of buildings include: eQuest (17%), EnergyPlus (12%), TRNSYS (8%), DOE2 (8%), DesignBuilder (6%), and Ecotect Analysis (5%). It should be noted that the general capabilities of airflow modelling in the above software packages are based on empirical equations for determining infiltration developed for residential buildings based on ASHRAE approaches and cannot be applied to taller buildings or buildings with natural or mechanical ventilation systems [61]. Therefore, a number of empirical methods for determining air exchange, which are reflected in software products, can be used for low-rise residential buildings [62–64].

In [75], proposed a method for estimating infiltration in commercial buildings using EnergyPlus [76], which takes into account wind effects, not just temperature effects. This approach is based on the use of constant coefficients A, B, C, and D for the DOE-2 and BLAST methods [75, 77]:

$$Infiltration = I_{design} \cdot F_{shedule}[\mathrm{A} + \mathrm{B} \cdot \Delta T + \mathrm{C} \cdot W + \mathrm{D} \cdot W^2], \tag{1.1}$$

where I_{design} is defined in EnergyPlus as the “design infiltration rate”, which is the air flow through the building envelope under design conditions; $F_{shedule}$ is a factor from 0.0 to 1.0 that can be set, usually to account for the effect of fan operation on infiltration [75, 77]; ΔT is the absolute difference in indoor and outdoor temperatures, °C; W is the wind speed in m/s. It should be noted that EnergyPlus changes the outdoor temperature and wind speed depending on the height of the zone for use in Eq. (1.1) and for other calculations. How this formula is used is described in detail in [75, 76]. A, B, C, and D are constants whose values are suggested in the EnergyPlus user manual [75, 76].

However, as mentioned above, this approach (the given coefficients) describes the air exchange processes in low-rise buildings with sufficient accuracy. In addition, the method provides a simplified view of the impact of building envelope permeability, weather, and HVAC system operation on the infiltration rate [28]. The method [28] does not take into account wind direction and number of storeys, which is especially important for leeward and windward facades of a building.

Among the software products for analysing air flows in buildings, CONTAM [28] is worth mentioning. CONTAM is a multi-zone computer program for air quality and ventilation analysis that allows determining air flows of infiltration, exfiltration and air flows into the room from controlled mechanical ventilation systems, wind pressures acting on the outer surface of the building, the stack effect caused by the difference in indoor and outdoor air temperatures, concentrations of pollutants and their impact on humans. It should be noted that this software product is used at the design stage and has not found its application in the energy management of existing buildings.

The software packages discussed above allow taking into account hourly changes in weather data based on typical year (TMY) data, for example, using the IWEC international climate file [78].

It should be noted that 80% of buildings in Ukraine meet the requirements of mass construction in the 1980s, which are characterised by a low level of energy efficiency, a trend that is also typical for post-Soviet countries [79, 80]. Assessment of the baseline energy consumption when implementing a set of energy saving measures is one of the key issues. In calculations of the baseline energy consumption, this value is taken in accordance with the normative values [76], and its actual value is difficult to determine experimentally, given that the natural component of this parameter is dynamic.

When calculating the energy performance of buildings, the stationary and quasi-stationary methods common in Ukraine use climatic characteristics for the relevant region/oblast centre, namely, the average monthly outdoor temperature, solar heat gain on horizontal and vertical surfaces for the quasi-stationary method, and the average for the heating period for the degree-day method [25]. The impact of wind in the models for monthly and seasonal averaging intervals is taken into account in a consolidated manner through the normative value of the air exchange rate. These methods do not allow analysing the change in energy demand, taking into account the time variability of operating conditions and external factors. For example, the normative climatological data in force in Ukraine provide monthly average values of the outdoor air temperature and solar heat gain on vertical and horizontal surfaces [81], which is sufficient for stationary and quasi-stationary calculation methods [81].

In the context of efficient energy use (heating regulation), the heating demand should be calculated taking into account hourly changes in climatic characteristics [25]. In Ukraine, dynamic models for determining the energy performance of buildings [25] are becoming increasingly used, requiring the use of hourly climate data, for example, hourly climate data for a typical year of the IWEC file [78]. Such approaches are relevant both for designing and deepening the analysis of the energy performance of existing buildings.

1.5 Standardisation of Indoor Climate Parameters

The issue of the quality of human living space is critical, as people spend a significant part of the day (almost 90% of the time) inside buildings: at home, at work or in educational institutions, without having information about the short-term and long-term health consequences of being in a particular room [82]. The quality of living space primarily depends on the quality of indoor air. The parameters of the air people breathe affect their health and comfort. The issue of indoor air quality is an important aspect of a healthy lifestyle, and its relevance has become particularly acute over the past few years, when, during the COVID-19 restrictions, people were forced to spend most of their day indoors, exposed to the pollutants accumulated there.

According to the World Health Organisation (WHO), indoor air pollution, both directly and indirectly, causes nearly 4 million premature deaths worldwide each year [83]. In particular, the negative effects of indoor air without proper ventilation can lead to symptoms such as discomfort, suffocation, difficulty breathing, headaches, drowsiness, and decreased mental and physical performance. A number of studies have been analysed on the monitoring and control of microclimate and air quality in schools [84], museums [85], universities [86], hospitals [87, 88], churches [89], aircraft cabin [90], subway platforms [91], and other facilities with relatively long stays of people during the day. Elevated concentrations of dust, nitrogen oxides, sulphur, carbon and formaldehyde can be detected in air samples, whether they are from the atmosphere or from indoor air.

The term "sick building syndrome" or SBS (sick building syndrome), proposed by the WHO, has been used to describe this phenomenon since 1982 [92]. The causes of sick building syndrome are related to two main factors: first, improved construction technologies and the fight for energy efficiency lead to the fact that building structures (walls, windows, doors and floors) become almost airtight, and second, synthetic materials are increasingly used in the decoration of premises and furniture production, which ultimately leads to an increase in emissions of harmful substances into the air of residential premises and makes it difficult to remove them [93].

To implement effective air quality monitoring, it is necessary to identify the main sources of air pollution. This will help to prepare the ground for the development of monitoring systems that can measure indoor pollutant concentrations and identify strategies for monitoring and improving indoor climate and air quality.

Indoor air pollution in living and working environments can be caused by a variety of occupant or employee activities, such as cooking, smoking, use of electronic devices, as well as emissions from building materials and the use of certain products. The harmful substances that can be present in the premises include carbon monoxide (CO), organic volatile compounds, particulate matter, aerosols, biological pollutants (mould, dust, bacteria) and others. It should be noted that the composition of indoor air varies significantly depending on the sources of pollution, emission rates and ventilation conditions.

A microclimate is a set of factors that directly affect the comfort of staying indoors [94]. Optimal microclimatic conditions are those microclimate parameters that, when exposed to a person for a long time and systematically, ensure a normal thermal state of the body without stress and disruption of thermoregulation mechanisms. In Ukraine, as in other countries of the world, there are a number of regulatory documents that define the requirements for buildings in terms of the microclimate in public and residential buildings. In addition, there are a number of documents that regulate the maximum permissible concentrations of harmful substances in the air and in the workplace.

- DSTU B EN 15,251:2011 "Calculated indoor climate parameters for the design and assessment of the energy performance of buildings in terms of air quality, thermal comfort, lighting and acoustics of buildings" [95]. This standard defines how to achieve and define the main parameters used as input data for energy calculation and long-term assessment of indoor climate, and defines the parameters used to monitor and display the indoor environment as recommended by the Energy Performance of Buildings Directive, to which different categories of criteria are applied depending on the type of building, type of occupants, type of climate and national characteristics.
- DSTU B EN 15,261:2012 "Calculation of microclimate parameters" defines the indoor microclimate parameters that affect the energy performance of buildings [96].
- DSTU B EN ISO 7730:2011 "Thermal Environment Ergonomics" establishes methods for predicting the overall thermal sensation and the degree of discomfort (thermal dissatisfaction) of people exposed to moderate thermal environments [97].
- State Standard 3.3.6.042–99 "Sanitary norms of microclimate of industrial premises" regulates the normative values of optimal and permissible microclimate indicators and sets requirements for methods of measuring microclimate parameters and their assessment [98].

It should be noted that according to these standards, the microclimate of residential and public buildings is determined mainly by the so-called thermal conditions—temperature, humidity and air movement, which determines its purity and freshness. These conditions are key aspects of air quality for two main reasons [99]: (1) many problems associated with poor microclimate can be solved by adjusting relative humidity or temperature, and (2) building materials in buildings can emit large amounts of pollutants into the air at high temperatures. All of these indoor climate parameters can not only be measured, but also controlled with climate control equipment. Regulatory documents define standards and methods for calculating indoor climate parameters to ensure comfortable living and working conditions.

According to these documents, the air temperature in the premises is regulated as follows. During the warm season, the recommended range is 22–25 °C. In the cold season, slightly lower: 20–23 °C for living rooms, 24–26 °C for bathrooms, 23–24 °C for children's rooms, and 18–20 °C for all other rooms [95, 96].

The recommended indoor humidity for humans is 40–60% [97]. Exceeding this mark is already dampness, which can lead to property damage and mould growth. Humidity below this level can have a negative impact on health (dry throat and eyes, dry skin).

Sanitary standards [98] specify the optimal humidity levels: 30–45% in winter and 30–60% in summer, and it should be borne in mind that children need more humid air than adults.

1.6 Pollutants in Indoor Air and Their Sources

It goes without saying that indoor air quality cannot be clearly separated from the quality of the air around us. The first factor affecting the microclimate is, of course, the level of air pollution in the vicinity of the building. Air from the street enters the building through windows, doors or the building's ventilation system. If the environmental conditions are unfavourable, the concentration of harmful substances indoors may also exceed the standard.

However, internal sources of air pollution in any room can have a much greater impact on the health and comfort of the person in it. In a new building, it is important to have the air analysed even before the renovation begins. Materials used in construction, such as concrete, can contain ammonia and other substances that are released from building structures over time and degrade the air quality in the home.

Substances such as formaldehyde, phenol, benzene, styrene, chromium, acetone, xylene and toluene can be released into the room after the renovation is complete. They can come from various building materials, varnishes, paints, adhesives and other substances used in the repair or manufacture of furniture. Naturally, the air quality in residential or office buildings is significantly affected by the quality of the outside air, indoor activities, as well as building and construction materials, equipment and furniture [99]. Some typical sources of indoor air pollution are shown in Fig. 1.5.

A review of studies was carried out [100–110]. On this basis, a list and analysis of the main pollutants, sources of their occurrence in the premises, and consequences for human health were made.

Fine particles in the air are carbon particles that adsorb organic chemicals and reactive metals. The most dangerous fraction of dust that can harm human health is fine dust less than 2.5 microns in size—the so-called $PM_{2.5}$ particles. The diameter of such particles is 20 times smaller than the diameter of a hair. Due to their small size, they are able to float freely in the air without settling on surfaces for a long

Fig. 1.5 Some typical sources of indoor air pollution

time. The WHO considers the maximum permissible concentration of $PM_{2.5}$ dust to be 25 μg/m^3 [100–111].

Sources of pollution: external (particles generated by the combustion of car fuel, tyre and asphalt abrasion, and can move from the outdoor environment to the indoor environment through windows, doors and ventilation systems) and internal (particles generated by internal activities such as cooking, smoking, machine operation and other household activities) [112, 113].

Consequences: particles can affect the lungs and heart, causing serious health problems.

Volatile organic compounds (VOCs) are gases that are released from liquids or solid materials. Advances in technology have led to an increase in the list of VOCs that are released into the air in homes and offices. This list includes aromatic hydrocarbons, esters, ketones, alcohols, acetates, as well as some of the most common VOCs—phenol and formaldehyde [104, 105].

Sources of pollution: plywood, chipboard and furniture made from them, paints, adhesives, rubbers, plastics, synthetic fabrics, tobacco smoke, etc. [106].

Consequences: The concentration of VOCs in indoor environments is usually at least 10 times higher than outdoors. They can cause irritation and health hazards, including cancer, with prolonged exposure.

Formaldehyde (CH_2O) is a toxic, allergenic and carcinogenic substance. It is classified as a hazard class 2 substance (highly hazardous). Despite its hazardousness, formaldehyde is often found in low concentrations in indoor air.

Sources of contamination: furniture, chipboard, polymeric materials, construction and finishing materials, cigarette smoke and emissions from domestic gas heating. Small amounts of formaldehyde are also found in exhaled air as a metabolic product.

Consequences: At sufficiently high concentrations, formaldehyde irritates the mucous membranes of the eyes, throat and respiratory tract, leading to headaches and nausea. The formaldehyde irritation threshold for the upper respiratory tract is 2.4 mg/m^3. Symptoms of formaldehyde poisoning can include pallor, weakness, fatigue, loss of consciousness, depression, difficulty breathing and sometimes night cramps. Chronic exposure to formaldehyde can lead to weakness, persistent headaches, weight loss, loss of appetite, heart rhythm disturbances and insomnia. Formaldehyde is particularly dangerous for children.

***Nitrogen oxides** (NOx)*

Pollution sources: The main nitrogen oxides are nitrogen oxide (NO) and nitrogen dioxide (NO_2), which are associated with combustion sources such as gas stoves and heaters.

Consequences: on average, the concentration of NO and NO_2 in buildings without combustion sources is half that of the outside, but in rooms with gas stoves and heaters, indoor levels often exceed the outdoor levels. NO is converted to NO_2 by oxidants, and therefore NO_2 is usually considered as the main pollutant [99, 105].

***Ozone** (O_3)*

Sources of pollution: O_3 is a strong oxidising agent mainly formed during photochemical reactions in the atmosphere. Indoor sources of ozone include the operation of electrical appliances such as photocopiers and air blowers [106].

Consequences: Ozone can interact with other chemical pollutants, which can cause irritation and damage to materials. The main routes of human exposure are inhalation and skin contact.

***Sulphur dioxide** (SO_2)*

Sources of pollution: SO_2 is the most abundant gas among the sulphur oxides (SOx) group in the atmosphere. The main sources of SO_2 emissions are the combustion processes of coal fuels and it mixes with aerosols and fine PM particles to form a complex group of air pollutants. Typically, SO_2 levels in buildings are lower than outdoors [107].

Consequences: may affect respiratory function.

Indoor ***carbon monoxide** (CO)* is mainly produced during combustion processes such as cooking or heating. In addition, CO can also enter the room through infiltration from the outdoors.

Consequences: High levels of CO can lead to negative health effects, including impacts on cardiovascular and neurobiological function, as well as causing a person to become unbalanced and even unexpectedly dead.

***Carbon dioxide** (CO_2),* a colourless and odourless gas, is a component of the Earth's atmosphere and also a major metabolite of humans. The concentration of CO_2 in ambient air is approximately 400 ppm and is mainly the result of the combustion of fossil fuels. The level of CO_2 in a room is used as an indicator of indoor air quality and for ventilation control. Indoor CO_2 concentrations should be below 700 ppm to ensure human health.

Consequences: Exposure to CO_2 has been found to cause headaches, drowsiness, fatigue, and difficulty concentrating [108].

Aerosols

Sources of pollution: primary aerosols (originating from various indoor and outdoor sources) and *secondary aerosols (*formed through the reactions of gases and particles in the room) [82].

Consequences: Effects on the respiratory system and effects on the heart and nervous system.

Toxic metals

Sources of pollution: internal (due to the use of certain devices and building materials, such as flooring, varnishes and paints) and external (through windows and ventilation) [109].

Consequences: risk of cancer, cardiovascular disease and other health problems.

Pesticides

Sources of pollution: internal (use of pesticides to protect wooden building materials in coatings) and external (air pollution from external pesticides) [110].

Consequences: short-term (skin and eye irritation, dizziness, headache, nausea) and long-term (risk of cancer, asthma and diabetes).

Radon is a radioactive inert gas formed during the decay of radium radionuclides. It has a half-life of about four days. It is believed that radon accounts for about 50% of the dose received by humans due to the natural radiation background. There is no safe concentration of radon [111].

Sources of pollution: construction materials (concrete, stone, brick). Ground gas and water sources. Radon emissions from ambient air.

Consequences: increased risk of lung cancer, in particular with prolonged exposure to high levels of radon.

Biological pollutants

Sources of pollution: biological allergens (animal hair, cockroaches, bedbugs, mites, pollen), microorganisms (viruses, fungi, bacteria) [112].

Consequences: respiratory infections, respiratory allergies, wheezing.

Table 1.7 summarises the main WHO guidelines for permissible concentrations of some common pollutants [113]. These standards are commonly used to monitor air quality inside households, schools, hospitals, public buildings and offices.

Indoor air quality has a major impact on the health and well-being of people who live or work there. Various sources of pollution, such as particles, volatile organic compounds, gases and biological agents, can lead to serious health effects, including respiratory illness, allergies, chronic diseases and even cancer.

1.7 Recommendations for Improving the Methodology of Indoor Climate Monitoring

Indoor air quality can be the subject of long-term measurements conducted by air pollution monitoring systems [114, 115]. Air pollution monitoring is an important component of indoor climate monitoring [116]. Before starting air pollution monitoring, it is necessary to identify the main sources and/or causes that may have the

Table 1.7 WHO recommendations for indoor air quality and monitoring of pollutant concentrations

Pollutants	Concentration (mg/m^3)	Permissible exposure time
CO	100 60 30 10	15 min 30 min 1 h 8 h
CO_2	1800	1 h
NO_2	0.4 0.15	1 h 24 h
PM (fine particles, dust)	0.15 0.05	24 h 1 year
O_3	0.15–0.2 0.1–0.12	1 h 8 h
SO_2	0.5 0.35	10 min 1 h
Lead	0.0005–0.001 0.0015	1 year 3 months
Formaldehyde	0.1	30 min
Radon	100 Bq/m^3	1 year

greatest impact on indoor air quality. The sources and causes that affect indoor air quality include: improperly installed ventilation [117]; chemical pollution accumulated over time [118]; pollution coming from outside into the room [119]; bacterial pollution [120]; pollution from building materials [101]; and other sources.

It is recommended to measure indoor air quality for the following substances: fine dust ($PM_{2.5}$ and PM_{10}), formaldehyde (CH_2 O), ozone (O_3), nitrogen dioxide (NO_2), sulphur dioxide (SO_2), carbon monoxide (CO) and carbon dioxide (CO_2) [121, 122]. These substances are the most common pollutants in the air and can significantly affect human health and well-being. Sources of their distribution in the premises and their impact on the body are described in Sect. 1.1.

Monitoring of indoor air pollution for selected pollutants should be carried out throughout the day with the ventilation system switched on to maintain comfortable conditions and prevent life-threatening situations. According to the WHO recommendations [83], the maximum permissible concentrations of these substances and the permissible time of exposure are as follows: PM—0.15 mg/m^3 (no more than 24 h); CH_2 O—0.1 mg/m^3 (no more than 30 min); O_3—0.15–0.2 mg/m^3 (no more than 1 h) and 0.1–0.12 mg/m^3 (no more than 8 h); NO_2—0.4 mg/m^3 (no more than 1 h) and 0.15 mg/m^3 (no more than 24 h); SO_2—0.5 mg/m^3 (no more than 10 min) and 0.35 mg/m^3 (no more than 1 h); CO—100 mg/m^3 (no more than 15 min); CO_2—1800 mg/m^3 (no more than 1 h).

The models of sensors and devices that can be used for air pollution monitoring and their selected characteristics proposed for the monitoring system sensor nomenclature are given in Tables A1 andA2 of Appendix A.

Sensors used for air pollution monitoring should provide a measurement interval sufficient to assess the reliability of the monitoring within the permissible exposure time of the relevant chemicals.

Assessment of the technical condition of the sensors (serviceability) is carried out using appropriate reference devices or calibration gas mixtures. In the absence of such devices and/or verification gas mixtures, the serviceability of the sensors of the measuring modules of the air pollution monitoring system is carried out by comparing the measurement results.

The comparability of sensor measurement results is carried out using the following statistical characteristics [123]:

– determination coefficient R^2

$$R^2 = \frac{\sum_{i=1}^{n} (\hat{y} - \bar{y})^2}{\sum_{i=1}^{n} (y_i - \bar{y})^2}, \tag{1.2}$$

where $\hat{y}$—estimated value of the regression; $\bar{y}$—average value of the experimental data; y_i—individual values of the experimental data;

– range R

$$R = |x_1 - x_2|, \tag{1.3}$$

where x_1—measurement result of the first sensor, x_2—measurement result of the second sensor;

– of the coefficient of variation C_v

$$C_v = \frac{S_x}{\bar{x}}, \tag{1.4}$$

where S_x—standard deviation, $\bar{x}$—average value of experimental measurements by a group of sensors.

The algorithm for assessing the technical condition of sensors using statistical characteristics is shown in Fig. 1.6.

If the determined statistical characteristics for each of the individual data groups are less than the set limit values (*k*, *l*, *m*, respectively), the sensor of the monitoring system can be characterised as faulty. If the obtained values of the statistical characteristics are greater than or equal to the set limit values, the sensor can be characterised as faulty.

The values of the determination coefficients R^2, which characterise the comparability of the data of the reference devices and PM, NO_2, CO and O_3 sensors obtained during laboratory (a) and field (b) experiments, are given in Appendix A in Figs. A1 and A3, respectively.

It has been experimentally established that the value of the limit coefficient of determination R^2 should not be less than 0.9 (coefficient *k*). The limit value of the range R (coefficient *l*), which indicates a sensor malfunction, is selected individually

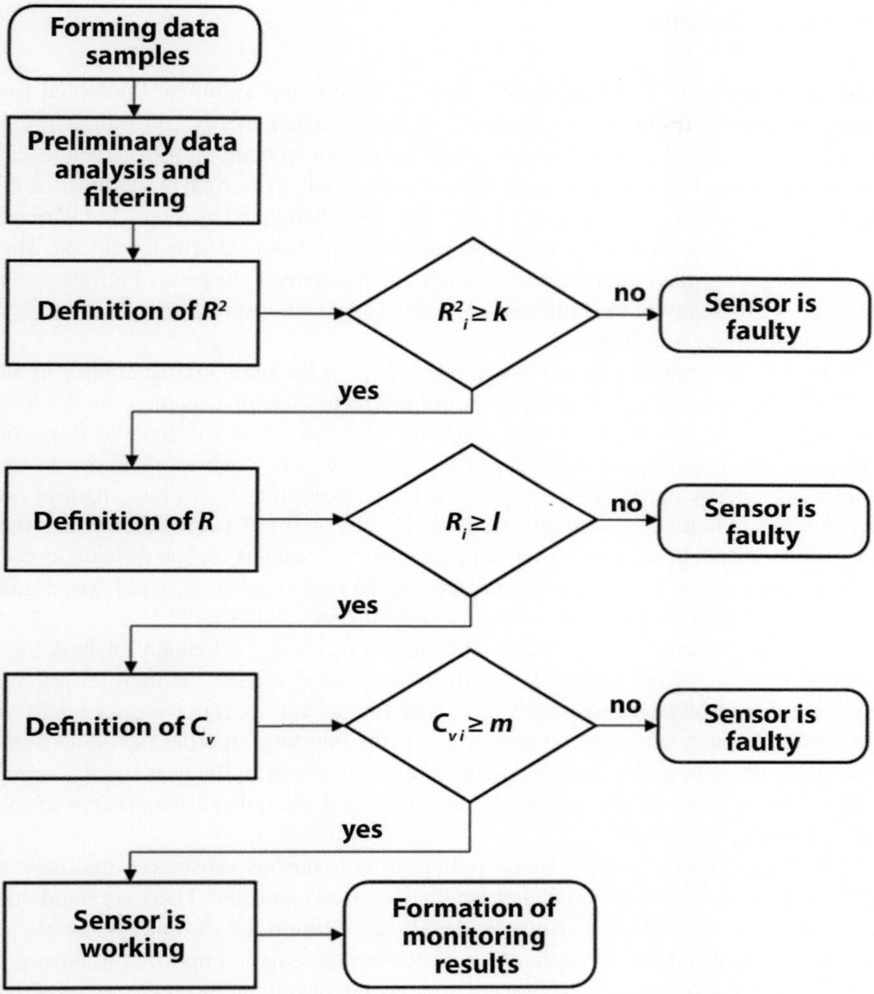

Fig. 1.6 for assessing the technical condition of air pollution monitoring sensors using statistical characteristics R^2, R, C_v

depending on the test conditions. A limit value of the coefficient of variation C_v (coefficient m) of more than 300% indicates a sensor malfunction (the sensor must be replaced or cleaned). The value of C_v in the range of 200–300% indicates the need for increased attention to this sensor.

1.8 Conclusions

The residential and public building sector in Ukraine has significant potential for energy efficiency improvements and is one of the largest energy consumers in the country, with energy efficiency significantly lower than in many European countries. In the EU and the US, important attention is paid not only to thermal modernisation of the building envelope, but also to the study of air exchange in buildings. In Ukraine, issues related to the study of air exchange rates in buildings need to be developed. The air exchange rate plays an important role not only in terms of the zone of influence on the energy efficiency of the building, but also in terms of ensuring indoor air quality and microclimate conditions.

Ensuring the standard air exchange rate is one of the main characteristics of an energy efficient building. Typically, the requirements for air exchange in residential premises are set for the entire dwelling, and less often for specific types of premises. The use of constant values of air exchange rates is advisable in the development of energy certificates and design. Under operating conditions, natural air exchange is dynamically changing, and indoor air quality depends not only on the air exchange rate, but also on other important factors, such as carbon dioxide levels, indoor and outdoor temperatures, air permeability of enclosures, wind speed and direction, number of storeys, number of occupants, etc.

The main informative indicators and models of energy efficiency of buildings are considered, it is established that in mathematical models of thermal regime and calculation of heat losses it is advisable to take into account weather data of the area, geometry and thermal properties of the building, internal heat emissions from lighting, people, equipment, technical characteristics of the heat supply, ventilation and cooling system, operation schedules and normalised parameters of the microclimate of premises.

Sources of indoor environmental pollution with various substances that have a potentially negative impact on human health have been identified. There are standards and regulations that define permissible levels of pollution for various substances in the indoor environment. Compliance with these standards is important to ensure a healthy indoor environment. Therefore, monitoring of indoor air quality is essential for the timely detection and elimination of pollution.

The most common ways to improve indoor air quality include choosing safer building materials, using smokeless heating systems, organising ventilation, etc.

References

1. On energy efficiency of buildings: Law of Ukraine of 22 June 2017 No. 2118-VIII. Verkhovna Rada of Ukraine. Bull. Verkhovna Rada Ukr. **33**, 359 (2024). https://ukrstat.gov.ua/operativ/operativ2020/zf/zf_reg/arh_zf_reg_u.htm

2. Concept of the state targeted economic programme for energy efficiency and development of renewable energy sources for 2021–2025, December 2020. https://saee.gov.ua/uk/pressroom/3612. Accessed 30 December 2023
3. Bilous, I.Y.: Assessment of energy efficiency of buildings in the conditions of dynamic change of environmental characteristics. Ph.D. thesis, speciality 05.14.01 "Energy systems and complexes", Igor Sikorsky Kyiv Polytechnic Institute, Kyiv (2019)
4. "European Environmental Agency. https://www.eea.europa.eu/data-and-maps/daviz/unit-consumption-of-space-heating#tab-chart_1. Accessed 29 Dec 2023
5. Residential buildings. Basic provisions, DBN B.2.2-15:2019. Ministry of Regional Development, Construction, Housing and Communal Services of Ukraine, Kyiv (2019)
6. Bilous, I.Y., Deshko, V.I., Sukhodub, I.O.: Building energy modelling using hourly infiltration rate. J. Civ. Eng. **4**(96), 27–41 (2020)
7. Guidelines for home energy professionals standard work specifications. Energy department. https://www.energy.gov/scep/guidelines-home-energy-professionals-standard-work-specifications. Accessed 1 Jan 2024
8. Energy Standard for Buildings Except Low-Rise Residential Buildings, ANSI/ASHRAE/IES Standard 90.1-2010, ANSI/ASHRAE (2012)
9. Order of the Ministry of Communities and Territories Development of Ukraine dated 27 October 2020 No. 261 "On Approval of Amendments to the Methodology for Determining the Energy Efficiency of Buildings", registered with the Ministry of Justice of Ukraine on 17 December 2020 under No. 1254/35537 (2020)
10. DSTU B EN 15459: Energy performance of buildings. Procedure for the energy assessment of building systems (EN 15459:2007, IDT) (2014)
11. Order of the Ministry of Communities and Territories Development of Ukraine dated 27 October 2020 No. 260 "On Approval of Minimum Requirements for Energy Efficiency of Buildings", registered with the Ministry of Justice of Ukraine on 18 December 2020 under No. 1257/35540 (2020)
12. DBN B.2.5-67: Heating, ventilation and air conditioning (2013)
13. DSTU-N B V.1.1-27: Building climatology, 130p. Ministry of the Region of Ukraine, Kyiv, 2011 (2010)
14. Order of the Ministry of Regional Development, Construction, Housing and Communal Services of Ukraine No. 169 "On Approval of the Methodology for Determining the Energy Efficiency of Buildings" dated 11 July 2018, registered with the Ministry of Justice of Ukraine on 16 July 2018 under No. 822/32274 (2018)
15. DBN B.2.6-31: Thermal insulation and energy efficiency of buildings, 33c. State Standard of Ukraine, Kyiv, 2022 (2021)
16. Antipov, E., Mishchenko, A., Shelimanova, O., Tarasenko, S.: Analysis of the influence of the internal heat capacity of the HEI building and weather-dependent regulation of the IHS on the efficiency of the heating system in standby mode. Energy Autom. **5** (2021). https://doi.org/10.31548/energiya2021.05.045
17. Buyak, N., Deshko, V., Bilous, I.: Changing energy and exergy comfort level after school thermomodernization. Rocznik Ochrona Srodowiska **23**, 458–469 (2021). https://doi.org/10.54740/ros.2021.031
18. Deshko, V., Buyak, N., Bilous, I., Voloshchuk, V.: Reference state and exergy based dynamics analysis of energy performance of the "heat source - human - building envelope". Energy **200** (2020). https://doi.org/10.1016/j.energy.2020.117534
19. Esmaeilzadeh, A., Zakerzadeh, M.R., Yousefi, K.A.: The comparison of some advanced control methods for energy optimisation and comfort management in buildings. Sustain. Cities Soc. **43**, 601–623 (2018). https://doi.org/10.1016/j.scs.2018.08.038
20. Nalyvaiko, V., Radko, I., Zhyltsov, A., Okushko, O., Mishchenko, A., Antypov, I.: Investigation of thermal modernised building's microclimate with renewable energy. In: ICoRES 2019. E3S Web of Conferences, vol. 154, p. 07011 (2020)
21. DSTU B EN ISO 7730: Erhonomics of the teplovoho seredovyshcha. Analitychne vyznachennia ta interpretatsiia teplovoho komfortu na osnovi rozrakhunkiv pokaznykiv PMV

i PPD i kryteriiv lokalnoho teplovoho komfortu [Accessed 2013-01-01], 74s. Kyiv: Minrehion Ukrainy, 2012 (2011)
22. DSTU B EN 15251: Rozrakhunkovi parametry mikroklimatu prymishchen dlia proektuvannia ta otsinky enerhetychnykh kharakterystyk budivela po vidnoshenniu do yakosti povitria, teplovoho komfortu, osvitlennia ta akustyky budivel [Accessed 2013-07-01], 71s. Ministry of Education of Ukraine, Kyiv, 2012 (2011)
23. DSTU B EN 15261: Rozrakhunok parametriv mikroklimatu [Kyiv, Ministry of Environment of Ukraine, 81s (2012)
24. Demchenko, V.V., Chupryna, H.M., Nevmerzhitsky, O.V.: Methods of increasing the energy efficiency of a building. Manag. Dev. Complex Syst. **16**, 138–143 (2013). http://nbuv.gov.ua/UJRN/Urss_2013_16_27.
25. German Association of Engineers, Calculation of transient thermal response of rooms and buildings - modelling of rooms. 91.140.10 (VDI 6007-1), Beuth Verlag GmbH, Düsseldorf (2012)
26. Reeves, T., Olbina, S., Raja, R.A.: ISSA guidelines for using building information modelling for energy analysis of buildings. Buildings **5**, 1361–1388 (2015)
27. TRNSYS energy simulation software. http://www.trnsys.com
28. Deshko, V., Sukhodub, I., Bilous, I.: Mathematical models for determination of specific energy need for heating used in Ukraine. J. New Technol. Environ. Sci. (JNTES) **1**, 13–25 (2018)
29. Fanger, P.: Assessment of man's thermal comfort in practice. Br. J. Ind. Med. **30**, 313–324 (1973)
30. Buyak, N., Deshko, V., Bilous, I., Biriukov, D., Voloshchuk, V.: Applying dynamic energy and exergy analysis to a building envelope. Int. J. Exergy **43**(2), 93–107 (2024). https://doi.org/10.1504/IJEX.2024.136928
31. Bilous, I., Deshko, V., Sukhodub, I.: Parametric analysis of external and internal factors influencing building energy performance using non-linear multivariate regression models. J. Build. Eng. **20**, 327–336 (2018)
32. Emmerich, S.J., Persily, A.K., Dowell, T.P.: Impact of infiltration on heating and cooling loads in US office buildings. In: Proceedings of the 26th IEA Conference of the Air Infiltration and Ventilation Centre, 21–23 September, Madison, pp. 1–7 (2005)
33. Biler, A., Tavil, A.U., Su, Y., Khan A.: Review of performance specifications and studies of trickle vent. Buildings **8**, 152–183 (2018)
34. Deshko, V., Bilous, I., Sukhodub, I., Yatsenko, O.: Evaluation of energy use for heating in residential building under the influence of air exchange modes. J. Build. Eng. **42**, 103020 (2021). https://doi.org/10.1016/j.jobe.2021.103020
35. Deshko, V.I., Bilous, I.Y., Hetmanchuk, G.O.: Study of air exchange in an apartment based on the experimental determination of CO2 mass transfer. Energ. Autom. **3**, 28–40 (2023)
36. Deshko, V., Bilous, I., Maksymenko, O.: Impact of heating system local control on energy consumption in apartment buildings. Rocznik Ochrona Srodowiska **25**, 77–85 (2023). https://doi.org/10.54740/ros.2023.009
37. Malyarenko, V.A., Orlova, N.A.: Analysis of the criterion of energy efficiency of buildings and structures. Integr. Technol. Energy Sav. **2**, 43–48 (2004)
38. Maslesa, I., Jensen, P., Birkved, M.: Indicators for quantifying environmental building performance: a systematic literature review. J. Build. Eng. **19**, 552–560 (2018)
39. Bilous, I., Deshko, V., Sukhodub, I.O.: Building inside air temperature parametric study. J. Civ. Eng. **8**, 65–75 (2016)
40. Dumała, S., Skwarczyński, M.: Influence of modernisation activities on thermal energy demand in buildings. Rocznik Ochrona Srodowiska **13**(1), 1795–1808 (2011)
41. Siuta-Olcha, A., Cholewa, T., Syroka, M., Anasiewicz, R.: Analysis of the influence of a glazed surface type and solar shading devices on the building energy balance. Rocznik Ochrona Srodowiska **18**, 259–270 (2016)
42. Deshko, V., Buyak, N., Bilous, I., Voloshchuk, V.: Reference state and exergy-based dynamics analysis of energy performance of the "heat source - human - building envelope" system. Energy (2020)

43. DSTU-N B V.3.2-3: Guidelines for the implementation of thermal modernisation of residential buildings [Effective from 2015-10-01]. Minregion of Ukraine, Kyiv, 2015, 40c. (2014)
44. Mechouet, A., Mouhib, T., Balhamri, A., Ouali, E.M.: Evaluating the impact of air infiltrations on the thermal and energy performances for different types of dwellings in casablanca city. Int. J. Mech. Prod. Eng. Res. Dev. (IJMPERD) **8**, 793–800 (2018)
45. Younes, C., AbiShihid, C., Bitsuamlak, G.: Air infiltration through building envelopes: a review. J. Build. Phys. **35–3**, 267–302 (2011)
46. Deshko, V.I., Bilous, I.Y., Hetmanchuk, G.O.: Calculation of the hourly natural air exchange rate in multi-storey buildings under conditions of variability of the external and internal environment. Coll. Sci. Works UkrDUZT **184**, 68–78 (2019)
47. Energy Performance of Buildings - Ventilation for Buildings - Part 1: Indoor Environmental Input Parameters for Design and Assessment of Energy Performance of Buildings Addressing Indoor Air Quality, Thermal Environment, Lighting and Acoustics - Module M1-6, EN 16798-1:2019, CEN, Brussels (2019)
48. Energy efficiency of buildings. Ventilation of buildings. Part 3. Ventilation in non-residential buildings. Operational requirements for ventilation and air conditioning systems (modules M5-1, M5-4) (EN 16798-3:2017, IDT), DSTU EN 16798-3:2019, State Enterprise "Ukrainian Research and Training Centre for Standardisation, Certification and Quality", Kyiv (2019)
49. Ventilation and Acceptable Indoor Air Quality, ASHRAE 62.1-2022, ANSI/ASHRAE, Atlanta, Georgia (2022)
50. Ventilation and acceptable indoor air quality in residential buildings, ASHRAE 62.2–2022, ANSI/ASHRAE, Atlanta, Georgia (2022)
51. Estimated parameters of indoor microclimate for the design and assessment of energy performance of buildings in relation to air quality, thermal comfort, lighting and acoustics (EN 15251:2007, IDT), DSTU B EN 15251:2011. State Enterprise "Ukrarkhbudinform", Kyiv (2012)
52. Energy efficiency of buildings. Method for calculating energy consumption during heating, cooling, ventilation, lighting and hot water supply, DSTU 9190:2022, SE NDIBK, Kyiv (2022)
53. Heating, ventilation and air conditioning, DBN V.2.5-67:2013. Ministry of Regional Development of Ukraine, Kyiv (2013)
54. Deshko, V.I., Bilous, I.Y., Sukhodub, I.O., Yatsenko, O.I.: Analysis of the impact of air exchange distribution between rooms on the energy consumption of an apartment. Energy Econ. Technol. Ecol. **1**(63), 39–50 (2021)
55. Buildings and structures. Educational institutions, DBN B.2.2-3-2018. Ukrainian Zonal Research and Design Institute for Civil Engineering (2018)
56. Almeida, R., Pinto, M., Pinho, P., Lemos, L.: Natural ventilation and indoor air quality in educational buildings: experimental assessment and improvement strategies. Energ. Effi. **10**, 839–854 (2017). https://doi.org/10.1007/s12053-016-9485-0
57. Song, Y., Sun, Y., Luo, S., Tian, Z., Hou, J., Kim, J., Dear, R.: Residential adaptive comfort in a humid continental climate - Tianjin China. Energy Build. **170**, 115–121 (2018). https://doi.org/10.1016/j.enbuild.2018.03.083
58. Albatayneh, A., Alterman, D., Page, A., Moghtaderi, B.: Thermal assessment of buildings based on occupants behaviour and the adaptive thermal comfort approach. Energy Procedia **115**, 265–271 (2017). https://doi.org/10.1016/j.egypro.2017.05.024
59. Prek, M.: Exergy analysis of thermal comfort. Int. J. Exergy **1**(Part. 2), 303–315 (2004)
60. Prek, M.: Thermodynamic analysis of human thermal comfort. Energy **31**(5), 732–743 (2006)
61. Prek, M., Butala, V.: Principles of exergy analysis of human heat and mass exchange with the indoor environment. Int. J. Heat Mass Transf. **53**(25–26), 5806–5814 (2010)
62. Shukuya, M.: Exergetic aspect of human thermal comfort and adaptation. In: Sustainable Houses and Living in the Hot-Humid Climates of Asia, pp. 123–129. Springer, Singapore (2018)
63. Juusela, M.A., Shukuya, M.: Human body energy consumption and thermal comfort of an office worker in typical and extreme weather conditions in Finland. Energy Build. **76**, 249–257 (2016)

64. Schweiker, M., et al.: Challenging the assumptions for thermal sensation scales, Build. Res. Inf. **45**(5), 572–589 (2017)
65. Isawa, K.: Human body energy balance: numerical analysis of an indoor thermal environment of a passive wooden room in summer. Buildings **5**, 1055–1069 (2015)
66. Deshko, V., Bilous, I., Vynogradov-Saltykov, V., Shovkaliuk, M., Hetmanchuk, H.: Integrated approaches to determination of co2 concentration and air rate exchange in educational institution. Rocznik Ochrona Srodowiska **22**(1), 82–104 (2020). https://ros.edu.pl/images/roczniki/2020/007_ROS_V22_R2020.pdf
67. Zhao, Z., Houchati, M., Beitelmal, A.: An energy efficiency assessment of the thermal comfort in an office building. Energy Procedia **134**, 885–893 (2017). https://doi.org/10.1016/j.egypro.2017.09.550
68. ASHRAE 55-2017: Thermal environmental conditions for human occupancy. American Society of Heating, Refrigerating and Air-Conditioning Engineers (ASHRAE), Atlanta (2017)
69. Humphreys, M., Nicol, F., Roaf, S.: Adaptive Thermal Comfort, Foundations and Analysis. Routledge, Earthscan, London (2015)
70. ASHRAE 554: Thermal environmental conditions for human occupancy (2004)
71. Haddad, S., Osmond, P., King, S.: Application of adaptive thermal comfort methods for Iranian schoolchildren. Build. Res. Inf. **47**(2), 173–189 (2019)
72. Rijal, H.B., Humphreys, M.A., Fergus Nicol, J.: Towards an adaptive model for thermal comfort in Japanese offices. Build. Res. Inf. 45(7), 717–729 (2017). https://doi.org/10.1080/09613218.2017.1288450
73. Impact of different thermal comfort models on zero energy residential buildings in hot climate. Energy Build. **102**, 117–128 (2015). https://doi.org/10.1016/j.enbuild.2015.05.017.
74. Fareniuk, H.H., Filonenko, O.I., Tymofieiev, M.V.: Enerhoefektyvnist hromadskykh budynkiv z vrakhuvanniam erhonomiky teplovoho seredovyshcha, Komunalne hospodarstvo mist. Seriia Technol. Sci. Archit. **135**, 119–124 (2017)
75. Deshko, V., Bilous, I., Buyak, N., Sapunov, A., Biriukov, D.: Dynamic interdependence of comfortable thermal conditions and energy efficiency increase in a nursery school building for heating and cooling period. Energy **283**, 129076 (2023). https://doi.org/10.1016/j.energy.2023.129076
76. EnergyPlus. https://energyplus.net/. Accessed 7 Dec 2024
77. Deshko, V., Bilous, I., Boiko, T., Shevchenko, O., Borodinecs, A., Zemitis, J.: Energy performance of higher education institutions buildings operating during quarantine restrictions and/or martial law in Ukraine. Latv. J. Phys. Techn. Sci. **61**(2), 44–65 (2024). https://doi.org/10.2478/lpts-2024-0013
78. International Weather for Energy Calculations. https://energyplus.net/weather-location/europe_wmo_region_6/UKR
79. Winkelmann, F.C.: Modelling windows in energyplus. In: Seventh International IBPSA Conference Riode Janeiro, Brazil, 13–15 August 2001. Building Simulation, pp. 457–464 (2001)
80. EN 15217: Energy performance of buildings - methods for expressing energy performance and for energy certification of buildings. European Committee for Standardisation,, CEN 20, 0731p. (2007)
81. Reeves, T., Olbina, S., Raja, R.A.: ISSA guidelines for using building information modelling for energy analysis of buildings. Buildings **5**, 1361–11388 (2015)
82. ,., Donatello, S., Cordella, M.: Level(s) indicator 4.1: indoor air quality user manual: introductory briefing, instructions and guidance (Publication version 1.1) (2021). https://susproc.jrc.ec.europa.eu/product-bureau/sites/default/files/2021-02/UM3_Indicator_4.1_v1.1_37pp.pdf
83. WHO indoor air quality guidelines: household fuel combustion. World Health Organisation (2014). https://www.who.int/news-room/fact-sheets/detail/household-air-pollution-and-health
84. Wysocka, M.: DALY indicator as an assessment tool for indoor air quality impact on human health. E3S Web Conf. **49**, 00133 (2018). https://doi.org/10.1051/e3sconf/20184900133

85. Ilieş, D.C., Marcu, F., Caciora, T., Indrie, L., Ilieş, A., Albu, A., Costea, M., Burtă, L.; Baias, Ş.: Investigations of museum indoor microclimate and air quality. Case study from Romania. Atmosphere **12**, 286 (2021). https://doi.org/10.3390/atmos12020286
86. Boeri, A., Longo, D., Fabbri, K., Pretelli, M., Bonora, A., Boulanger, S.: Library Indoor microclimate monitoring with and without heating system. A bologna university library case study. J. Cult. Herit. **53**, 143–153 (2022). https://doi.org/10.1016/j.culher.2021.11.012
87. Cannistraro, G., Bernardo, E.: Monitoring of the indoor microclimate in hospital environments a case study the Papardo hospital in Messina. Int. J. Heat Technol. **35**, S456–S465 (2017). https://doi.org/10.18280/ijht.35Sp0162
88. Leung, M., Chan, A.H.: Control and management of hospital indoor air quality. Med. Sci. Monit. **12**(3), SR17–SR23 (2006). Epub 2006 Feb 23. PMID: 16501436
89. Wysocka, M.: Analysis of indoor air quality in a naturally ventilated church. E3S Web Conf. **49**, 00134 (2018).s https://doi.org/10.1051/e3sconf/20184900134
90. Boschi, N., Haghighat, F.: Aircraft cabin indoor air environment requirements. In: Hocking, M. (eds) Air Quality in Aircraft Cabins and Similar Enclosed Spaces. The Handbook of Environmental Chemistry, vol. 4H. Springer, Berlin, Heidelberg. https://doi.org/10.1007/b107236
91. Kim, M.J., Braatz, R.D., Kim, J.T., Yoo, C.K.: Indoor air quality control for improving passenger health in subway platforms using an outdoor air quality dependent ventilation system. Build. Environ. **92**, 407–417 (2015). https://doi.org/10.1016/j.buildenv.2015.05.010, ISSN 0360-1323
92. Passarelli, G.R.: Sick building syndrome: an overview to raise awareness. J. Build. Apprais. **5**, 55–66 (2009). https://doi.org/10.1057/jba.2009.20
93. Shahzad, S.S., Brennan, J., Theodossopoulos, D., Hughes, B., Calautit, J.K.: Building-related symptoms, energy, and thermal control in the workplace: personal and open plan offices. Sustainability **8**(4), 331. https://doi.org/10.3390/su8040331
94. The microclimate in the house: parameters, requirements and control. https://navyflex.com.ua/en/the-microclimate-in-the-house-parameters-requirements-and-control/
95. DSTU B EN 15251: Design parameters of the indoor microclimate for the design and assessment of the energy performance of buildings in relation to air quality, thermal comfort, lighting and acoustics of buildings [Effective from 2013–07–01]. Kyiv: Ministry of Regional Development of Ukraine, 2012, 71c (2011)
96. DSTU B EN 15261: Calculation of microclimate parameters [Effective from 2013-01-01]. Kyiv: Ministry of Regional Development of Ukraine, 81c (2012)
97. DSTU B EN ISO 7730: Ergonomics of the thermal environment. Analytical determination and interpretation of thermal comfort based on the calculation of PMV and PPD and local thermal comfort criteria [Effective from 2013-01-01]. Kyiv: Ministry of Regional Development of Ukraine, 2012, 74c (2011)
98. DSTU 3.3.6.042-99: Sanitary norms of microclimate of industrial premises. https://ips.ligazakon.net/document/MOZ642
99. What are the main indoor air pollutants. https://vencon.ua/ua/articles/kakovy-istochniki-zagryazneniya-vozduha-v-pomeshchenii
100. Occupational Safety and Health Administration: Technical manual: indoor air quality investigation. https://www.osha.gov/dts/osta/otm/otm_iii/otm_iii_2.html
101. Tran, V.V., Park, D., Lee, Y.-C.: Indoor air pollution, related human diseases, and recent trends in the control and improvement of indoor air quality. Int. J. Environ. Res. Public Health **17**, 2927 (2020). https://doi.org/10.3390/ijerph17082927
102. Leung, D.Y.C.: Outdoor-indoor air pollution in urban environment: challenges and opportunity. Front. Environ. Sci. **2**, 69 (2015)
103. Environmental Protection Agency. Indoor Air Quality (IAQ): Indoor particulate matter indoor particulate matter|US EPA
104. Li, A.J., Pal, V.K., Kannan, K.: A review of environmental occurrence, toxicity, biotransformation and biomonitoring of volatile organic compounds. Environ. Chem. Ecotoxicol. **3**, 91–116 (2021). https://doi.org/10.1016/j.enceco.2021.01.001

105. Kodama, Y., Arashidani, K., Tokui, N., Kawamoto, T., Matsuno, K., Kunugita, N., Minakawa, N.: Environmental NO2 concentration and exposure in daily life along main roads in tokyo. Environ. Res. **89**, 236–244 (2002). https://doi.org/10.1006/enrs.2002.4350
106. Weschler, C.J.: Ozone in indoor environments: concentration and chemistry. Indoor Air **10**, 269–288 (2000). https://doi.org/10.1034/j.1600-0668.2000.010004269.x
107. Seow, W.J., Downward, G.S., Wei, H., Rothman, N., Reiss, B., Xu, J., Bassig, B.A., Li, J., He, J., Hosgood, H.D., et al.: Indoor concentrations of nitrogen dioxide and sulfur dioxide from burning solid fuels for cooking and heating in Yunnan Province, China. Indoor Air **26**, 776–783 (2016). https://doi.org/10.1111/ina.12251
108. Babak, V., Zaporozhets, A., Kuts, Y., Fryz, M., Scherbak, L. (2025). Identification of Air Pollution Sources. In: Noise signals. Studies in Systems, Decision and Control, vol 567. Springer, Cham. https://doi.org/10.1007/978-3-031-71093-3_7
109. Morawska, L., Afshari, A., Bae, G.N., Buonanno, G., Chao, C.Y.H., Hänninen, O., Hofmann, W., Isaxon, C., Jayaratne, E.R., Pasanen, P., et al.: Indoor aerosols: from personal exposure to risk assessment. Indoor Air **23**, 462–487 (2013). https://doi.org/10.1111/ina.12044
110. Tan, S.Y., Praveena, S.M., Abidin, E.Z., Cheema, M.S.: A review of heavy metals in indoor dust and its human health-risk implications. Rev. Environ. Health **31**, 447–456 (2016). https://doi.org/10.1515/reveh-2016-0026
111. USEPA: Pesticides' impact on indoor air quality. https://www.epa.gov/indoor-air-quality/iaq/pesticides-impact-indoor-air-quality
112. WHO: Who handbook on indoor radon: a public health perspective. World Health Organisation, Geneva, Switzerland (2009)
113. US Environmental Protection Agency: Indoor Air Quality (IAQ): What are biological pollutants, how do they affect indoor air quality? https://www.epa.gov/indoor-air-quality-iaq/what-are-biological-pollutants-how-do-they-affect-indoor-air-quality
114. Wolkoff, P.: Indoor air humidity, air quality, and health-an overview. Int. J. Hyg. Environ. Health **221**(3), 376–390 (2018)
115. Burroughs, H.E., Hansen, S.J.: Managing Indoor Air Quality. River Publishers (2020)
116. Kalva, O.: The role of microclimate in the formation of indoor air pollution. Therm. Sci. **27**, 105 (2023)
117. Ruan, T., Rim, D.: Indoor air pollution in office buildings in mega-cities: effects of filtration efficiency and outdoor air ventilation rates. Sustain. Cities Soc. **49**, 101609 (2019)
118. Gonzalez-Martin, J., Kraakman, N.J.R., Perez, C., Lebrero, R., Munoz, R.: A state-of-the-art review on indoor air pollution and strategies for indoor air pollution control. Chemosphere **262**, 128376 (2021)
119. Balmes, J.R.: Household air pollution from domestic combustion of solid fuels and health. J. Allergy Clin. Immunol. **143**(6), 1979–1987 (2019)
120. Liu, H., Zhang, X., Zhang, H., Yao, X., Zhou, M., Wang, J., He, Z., Zhang, H., Lou, L., Mao, W., Zheng, P., Hu, B.: Effect of air pollution on the total bacteria and pathogenic bacteria in different sizes of particulate matter. Environ. Pollut. **233**, 483–493 (2018)
121. Vardoulakis, S., Giagloglou, E., Steinle, S., Davis, A., Sleeuwenhoek, A., Galea, K.S., Dixon, K., Crawford, J.O.: Indoor exposure to selected air pollutants in the home environment: a systematic review. Int. J. Environ. Res. Public Health **17**(23), 8972 (2020)
122. Babak V.P., Babak S.V., Eremenko V.S., Kuts Y.V., Myslovych M.V., Scherbak L.M., Zaporozhets A.O. Models and Measures for Atmospheric Pollution Monitoring // Models and Measures in Measurements and Monitoring. - 2021. - vol. 360. - pp. 227-266. https://doi.org/10.1007/978-3-030-70783-5_8
123. Babak, V., Babak, S., Eremenko, V., Kuts, Y., Zaporozhets, A. (2025). Quantitative Evaluations of Measurement Information. In: Information-Measuring Systems. Studies in Systems, Decision and Control, vol 592. Springer, Cham.https://doi.org/10.1007/978-3-031-89406-0_2

Chapter 2
Models and Tools for Analyzing the Thermal Conditions of Buildings

2.1 Building Thermal Model

In accordance with the methodology of system analysis, the model of the thermal regime of a building as a single heat and power system should be presented in the form of three interrelated models that are more convenient for study (Fig. 2.1).

Taking into account the accepted division of the model of the thermal regime of a building as a single heat and power system, it can be written down into three interdependent submodels:

$$Q_{tr} = \eta_1 \cdot \eta_2 \cdot \eta_3, \tag{2.1}$$

where η_1 is the heat and energy efficiency indicator of the impact of external climatic parameters (temperature, wind speed and direction, level of solar radiation) on the building; η_2 is the heat and energy efficiency indicator of the optimal choice of heat and sun protection characteristics of external enclosing structures, as well as their massiveness; η_3 is the heat and energy efficiency indicator of the optimal choice of microclimate systems and the level of their automation.

A model of the thermal energy impact of the external climate on a building (η_1). This is a combination of climatic parameters, internal heat emissions and absorptions, and the thermal protection properties of the envelope. When calculating the heat transfer through the envelope, the effect of these factors must be taken into account together: for winter, the determining climate parameters, for example, are the outside air temperature $t_{3.n}$, and wind speed v. In some calculations, the relative humidity φ and enthalpy *i of* the outside air, as well as solar radiation and wind direction, should be taken into account in addition to these factors, adjusted for the height of the building [1].

The harmonious nature of the change in climatic parameters (temperature and enthalpy of the outside air, intensity of solar radiation) allows them to be determined by the function of the season in the form of a trigonometric series. At the same time,

A. Zaporozhets et al., *Buildings Energy Audit*, Studies in Systems, Decision and Control 641, https://doi.org/10.1007/978-3-032-07200-9_2

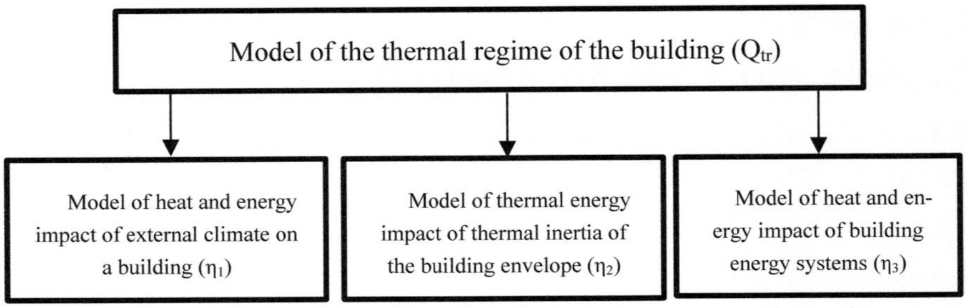

Fig. 2.1 Model of the building thermal regime

sufficient accuracy is achieved by the simplest approximation of the first two terms of the series, which is the limited consideration here [2]:

$$y = y_y + A_y cos\overline{z}, \tag{2.2}$$

where y_y—average annual value of the parameter; A_y—annual amplitude of its change; $\overline{z}$—relative time of annual change of the parameter

$$\overline{z} = \frac{z}{365} 2\pi, \tag{2.3}$$

where $z = z' - z^{max}$—local time, days, from the moment the parameter reaches the maximum; z'—also, counted from 1 January; z^{max}—time when the parameter reaches the maximum value, counted from 1 January.

The values of y_y, A_y, z^{max} for different climatic parameters and geographical areas can be obtained by processing observation data [1].

Using formula (2.2), similarly to formula (2.4), it is possible to find the duration Δz *of* the climate parameters above a given value in a year as the inverse trigonometric function of this value [2]:

$$\Delta z = \frac{365}{\pi} \arccos\big(y - y_y\big)/A_y. \tag{2.4}$$

However, in some cases, it becomes necessary to determine the duration of the *y* parameter when its value goes beyond the monthly average. In this case, the variability of the entire set of values of the parameter can be determined using statistical functions of the distribution of these values during the year.

Reference [1] shows the temperature and enthalpy of air, which are close to a sinusoid by the formula. Therefore, within the limits where such an approximation is valid, the sine wave has an amplitude equal to 365/2 days, and a quarter of its period Δy_{out} is equal to the change in the parameter from the annual average $y_{out.a}$ to its extreme (maximum) $\Delta y_{.extr.max}$ or minimum $\Delta y_{.extr.min}$ value. It should be particularly noted that, according to the law of sine, in this case, the time z changes as the parameter y_{out} changes with a period equal to $4\Delta y_{out}$:

$$z = \frac{365}{2} - \frac{365}{2}\sin\frac{2\pi}{4\Delta y_{out}}(y_{out} - y_{out.a}), \tag{2.5}$$

where $\Delta y_{out} = (\Delta y_{.extr} - y_{out.a})$—quarter of the period; $y_{out.a}$—average annual value of the parameter; $y_{.extr}$—the same, extreme (for the maximum and minimum values).

By marking

$$\overline{y_{out}} = \frac{y_{out} - y_{out.a}}{\Delta y_{out}}, \tag{2.6}$$

and expressing Δz in hours, we will obtain an approximate analytical function of the duration of the external climate parameter values above the one given y_{out}:

$$\Delta z = 4.38 \cdot 10^3\left(-sin\frac{\pi}{2}\overline{y_{out}}\right). \tag{2.7}$$

To determine the duration of the ambient air parameters below the set point, change the minus sign in formula (2.7) to a plus sign.

If there are data on the probable duration of standing Δz_{yout} of different values of the external climate parameters, the values of Δu_{out} can be determined by the formula

$$\Delta y_{out} = \frac{\sum y_{out.a}^{out.extr}(y_{out} - y_{out.a})\Delta z_{y_{out}}}{66.5}. \tag{2.8}$$

Taking into account the logical connection between the annual course of changes in monthly mean values (2.2) and the function of the annual distribution of the climate parameter (2.5), an approximate relationship for determining the values of Δy_{out} can be recommended:

$$\Delta y_{out} \approx 2A_{y_{out}}, \tag{2.9}$$

where $A_{y_{out}}$—amplitude of the annual variation of the average monthly values of the parameter y_o.

The values of Δz and $\overline{\Delta z}$ (в $\overline{\Delta z}$ value Δz, h/year, related to 365 – 24 = 8760 h/year) can be determined using the nomogram [2].

Thus, taking into account the harmonious nature of the annual change in the factors affecting heat exchange in the room, the thermal regime of the room can be represented as follows

$$Q_r = Q_{r.a} + A_{Q_r}cos\bar{z}, \tag{2.10}$$

where $Q_{r.a}$—average value (during the year) of the harmoniously changing thermal condition of the room (excess or lack of heat); A_{Q_r}—amplitude of annual deviation of the thermal condition of the room from the average.

For the sake of simplicity, we will assume that the annual change in indoor air temperature approaches a harmonic oscillation. In accordance with this

$$t_{in} = t_{in.a} + A_{t_{in}} cos\overline{z}. \tag{2.11}$$

The average annual value of the internal air temperature $t_{in.a}$ is defined as the average between the winter $t_{in.winter}$ and summer $t_{in.summer}$ calculated values, and the amplitude $A_{t_{in}}$ is equal to half of their difference.

Considering that during the heating period the heat balance of the room becomes negative (lack of heat), i.e. $\overline{Q}_r$ solving Eq. (2.10), we can determine the duration of the heating period Δz_h, days, as an inverse trigonometric function [2]:

$$\Delta z_h = \frac{365}{\pi} arccos\overline{Q}_r = 116 arccos\overline{Q}_r, \tag{2.12}$$

where $\overline{Q}_r$—dimensionless value

$$\overline{Q}_r = (Q_r - Q_{r.a})/A_{Q_r}. \tag{2.13}$$

Formula (2.13) is general in nature. It can also be used to calculate Δz_{hp} for the central control of the heating system from t_3. Since the duration of the heating period is defined as a part of the year with $t_{out} < 8\,°C$, formula (2.14) takes the form

$$\bar{t}_{out} = (t_{out.a} - 8)/)A_{t_{out.a}}. \tag{2.14}$$

At the same time, the values of Δz_h values most closely correspond to those given in [1].

The annual heat consumption for heating can be represented as the integral sum of the thermal state of a room changing during the heating period [2]:

$$Q_{h.a} = \int Q_r dz = 0.143 mn' A_{Q_r}\left(116 sin\frac{1}{2}\Delta\overline{z} - \overline{Q}_r \Delta z_h\right), \tag{2.15}$$

where m is the number of hours of heating system operation per day; n' is the number of days of heating system operation per week.

In the usual interval of change $\overline{Q}_r$ (from −0.6 to +0.6), it is convenient to use an approximate formula that gives a fairly accurate result [2]:

$$Q_{h.a} = 0.143 mn' A_{Q_r}\left[116 - \overline{Q}_r(91.2 + 0.5\Delta z_h)\right]. \tag{2.16}$$

In [2], the dependence curve $K_h = f(\overline{Q}_п)$ constructed by the formula

$$K_h = 116 sin\frac{1}{2}\Delta\overline{z} - \overline{Q}_r \Delta z_h. \tag{2.17}$$

Taking this into account, formula (2.15) takes the form

$$Q_{h.a} = 0.143143mn'A_{Q_r}K_h. \tag{2.18}$$

The calculation of the heat consumption for ventilation and the duration of its operation can be performed similarly to that for heating.

Model of the thermal energy impact of thermal inertia of the building envelope (η_2). Here, special attention, of course, should be paid to the thermal modernisation of buildings in order to bring the existing level of heat transfer resistance of external envelope structures to the standardised values R_{qmin}, which are defined by the provisions of the current standards [3] and fully meet the requirements of Table B.2 of Appendix B. Thus, taking into account the above, the indicator of thermal energy efficiency in terms of the optimal choice of thermal and sun protection characteristics of external envelope structures is $\eta_2 = 1$, provided that

$$R_{\Sigma rh} \geq R_{qmin}, \tag{2.19}$$

where $R_{\Sigma rh}$ is the reduced heat transfer resistance of an opaque or translucent building envelope, m^2·K/W.

The indicator of thermal energy efficiency in terms of the effect of the massiveness of the building envelope will be considered in conjunction with the following model, taking into account the values of outdoor air temperatures, depending on the heat capacity of the building, determined according to [4], which are given in Table B.2 of Annex B.

Model of thermal energy impact of building engineering systems (η_3). The efficiency of heat energy use by the heating system is presented in the international standard [5].

Similar to the efficiency method adapted from the German standard DIN 18,599, the approach is used in the national standard DSTU 9190:2022 "Energy Efficiency of Buildings. Method for calculating energy consumption for heating, cooling, ventilation, lighting and hot water supply" [4].

The calculated heat energy consumption by the heating system for creating a microclimate in the premises of a building $Q_{em,ls,year}$, for the heating period, depending on the degree of detail of the factors affecting the energy efficiency of the system [4, 6]—equipment, circuit design, control means, characteristics of the heated room, is determined by equation:

$$Q_{em,ls,year} = \frac{f_{hydr} \cdot f_{im} \cdot f_{rad}}{\eta_{em}} \cdot \sum_{i-1}^{n} [Q_k - (Q_{in} + Q_s)\nu], \tag{2.20}$$

where f_{hydr} is a coefficient that takes into account the hydraulic balancing of the system; f_{im} is a coefficient that takes into account the use of a periodic thermal regime of the room; f_{rad} is a coefficient that takes into account the effect of radiant heat transfer; η_{em} is a generalising coefficient that takes into account the heat transfer conditions of the system and is determined by the formula:

$$\eta_{em} = \frac{1}{(4 - \eta_{str} - \eta_{ctr} - \eta_{rad})}, \tag{2.21}$$

where η_{str}—coefficient that takes into account the effect of the gradient (stratification) of the indoor air temperature; for some systems, it is the arithmetic mean of the following coefficients: η_{str1}—takes into account the temperature of the heat carrier and η_{str2}—takes into account the installation conditions of the heater; η_{ctr}—coefficient that takes into account the type of indoor air temperature control used; η_{emb}—coefficient that takes into account the heat input to the heated room from built-in heating elements (for panel radiant systems); for some systems, it is the arithmetic mean of the following coefficients: η_{emb1}—takes into account the type of panel-radiant system and η_{emb2}—takes into account the thermal insulation of the panel-radiant system to adjacent rooms; n—number of full and incomplete ith months of the heating period; Q_k—total heat losses of the building through the building envelope in the ith month of the heating period, kWh.; Q_{in}—internal heat gain in the ith month of the heating period, kWh; Q_s—heat gain through windows and other translucent building envelopes from total solar radiation (direct and diffuse) under average cloudy conditions in the ith month of the heating period, kWh; v—heat gain utilization rate:

- for buildings without automatic indoor air temperature control $v = 0$;
- for buildings with automatic indoor air temperature control is determined in accordance with Fig. 2.2 [7], according to the thermal inertia criterion D, which is determined according to the algorithm given in DSTU-N B V.2.6–190.

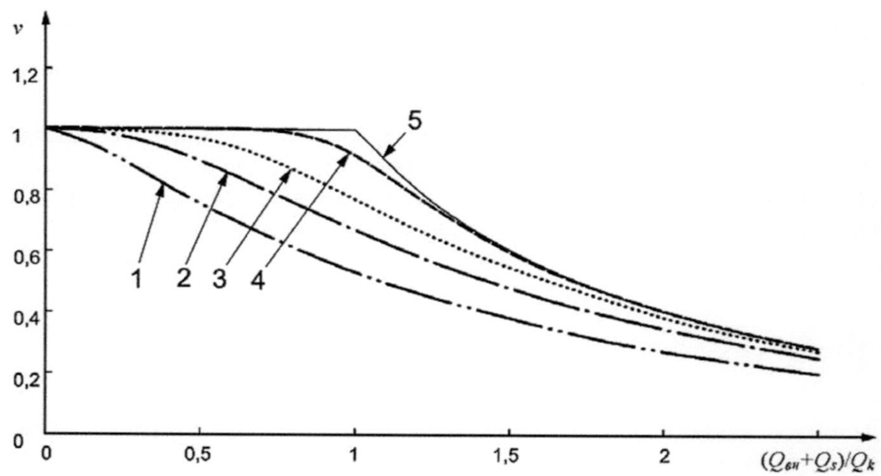

Fig. 2.2 Heat gain utilisation factor [7] for buildings with thermal inertia D: 1 – D ≤ 1.5; 2 – D = 2.3; 3 – D = 3.2; 4 – D ≥ 4; 5 – ideal curve

During performing calculations to compare the efficiency of balancing single-pipe heating systems of a building and assess its impact on heat energy savings in a building of a higher education institution (HEI), formulas (2.1)–(2.21) were adopted:

- a building of a higher education institution, with a water heating system with heating devices (radiator, convector, etc.) and rooms up to 4 m high;
- the heating devices are equipped with thermostats with P-regulation (2 K), the temperature head is 60 K (for example, 90/70);
- heating devices are installed under windows with radiation protection (see Table 17 [4]);
- temperature zone of the building—I;
- all heat transfer resistances of external envelope structures meet condition (2.20) and satisfy the current standards [3].

The coefficients to formulas (2.20) and (2.21), depending on energy efficiency factors, are presented in Tables B.3 and B.4 of Appendix B.

Table B.4 of Appendix B shows the heat energy savings obtained by applying different automatic balancing options—for a single-pipe system with a constant hydraulic mode and separately for a single-pipe heating system with a variable hydraulic mode. Unbalanced systems were used as the basic comparison options.

Below, in Figs. 2.3 and 2.4, graphical dependencies are presented that take into account the impact of the internal heat capacity of the building, determined according to [4], and the methods of balancing the heating system on the thermal regime and heat energy savings of the HEI building. A single-pipe unadjusted system is taken as the basic comparison option.

Based on the analysis of the results shown in Fig. 2.3, it can be argued that as the value of the internal heat capacity of a building increases, the reduction in specific

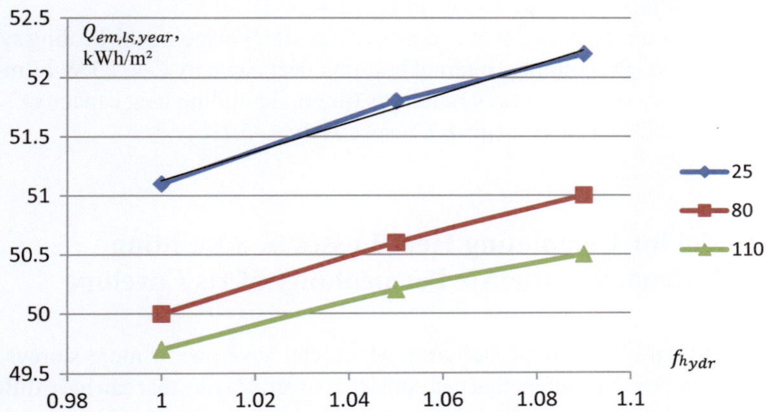

Fig. 2.3 Influence of the internal heat capacity of the building, determined according to [4], and methods of balancing the heating system on the thermal regime of the HEI building (assuming that $K_{out(PD)} = 1.0$), W·h/(m²·K): —C = 25; —C = 80; —C = 110

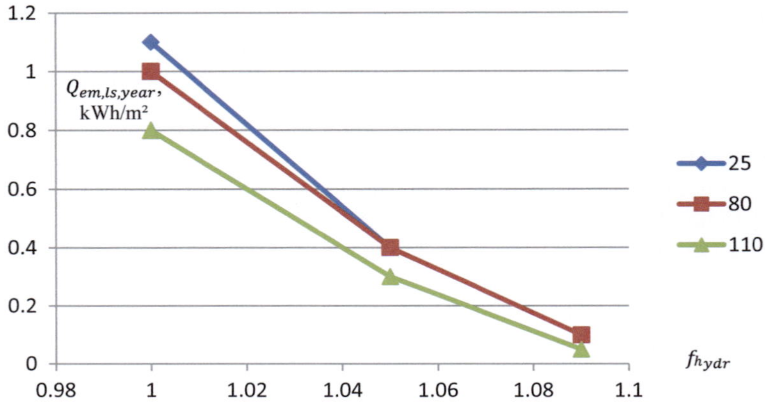

Fig. 2.4 Influence of the internal heat capacity of the building, determined according to [4], and methods of balancing the heating system on the heat energy saving by the HEI building (assuming that $K_{out(PD)} = 1.0$), W·h/(m^2·K): —C = 25; —C = 80; —C = 110

energy consumption by a HEI building becomes less noticeable compared to less massive buildings (light and very light construction). For example, for buildings with an internal heat capacity of C = 110 W·h/(m^2·K), the specific indicator of energy consumption reduction by a HEI building is 0.8 thousand kWh./m^2, while for dormitories with an internal building heat capacity of C = 80 W·h/(m^2·K) and C = 25 W·h/(m^2·K), the specific reduction in energy consumption by the building was 1.0 thousand kWh/m^2 and 1.1 thousand kWh/m^2 per year, respectively.

This is illustrated by the graphs of the impact of the internal heat capacity of the building and the methods of balancing the heating system on the heat energy savings of the HEI building, which are shown in Fig. 2.4.

In percentage terms, the level of energy savings after balancing the heating system for a very lightweight building (internal building heat capacity C = 25 W·h/(m^2·K)) was 2%, for heavy and very heavy buildings (internal building heat capacity C = 80 W·h/(m^2·K) and C = 110 W·h/(m^2·K), respectively)—2.2%.

2.2 Model for Calculating Heat Losses in a Building with Regard to the Air Permeability of Its Envelope

The object of study is a single building, which can have one or more storeys. The building envelope is a structure that can consist of several layers that can have different thermal properties, such as thermal conductivity, heat capacity, air permeability, moisture permeability, heat storage properties, etc. The building has N_d doors, with the central entrance door being a double door. There are N_w windows built into the wall structures, where air penetration (infiltration) is possible at the joints with the wall. Two types of rooms are considered, with or without ventilation openings. The building is heated using n_b batteries, each of which has a surface area F_b and a surface temperature T_b. It is assumed that the outside air may have a temperature $T_o\ (t)$, which varies throughout the year in winter and summer, where t *is* a time variable. It should be noted that this temperature also depends on the region in which the building is located, since the temperature characteristics throughout the year will differ for different regions. It is assumed that the wind speed flowing on the building surface is W_{ws}, and its value can vary depending on the weather and building density in cities with a large number of buildings. In addition, there is external thermal radiation in the external environment, the value of which varies depending on cloud cover.

The heat balance equation, which takes into account all components of heat gain from heat sources inside the building and heat loss from the building to the environment, is as follows:

$$Q_b + Q_{ll} + Q_{ea} + Q_p + Q_{sr} = Q_{wall} + Q_w + Q_d + Q_f + Q_c + Q_v + Q_{ap}, \quad (2.22)$$

where Q_b is the amount of heat entering the room from heating batteries, J; Q_{ll} is the amount of heat emitted from lighting lamps, J; Q_{ea} is the heat input from household electrical appliances located in the building, J; Q_p is the heat input from people in the room, Q_{sr} is the heat input to the room from solar radiation, J; Q_{wall}—heat losses to the environment from the building envelope, J; Q_w—heat losses through windows, J; Q_d—heat losses through doors, J; Q_f—heat losses through the floor, J; Q_c—heat losses through the ceiling, J; Q_v—heat losses through ventilation openings, J; Q_{ap}—heat losses due to air permeability through the wall envelope, J.

Let us consider the methods of calculating each component included in Eq. (2.22). The amount of heat transferred to the internal air from the batteries is found from the equation:

$$Q_b = n_b \alpha_b F_b (T_b - T_{in}), \quad (2.23)$$

where n_b—number of batteries in the room, *pcs*; α_b—heat transfer coefficient on the battery surface, W/m°C; F_b—battery surface area, m^2; T_b—temperature on the surface of the battery, °C; T_{in}—indoor air temperature, °C.

The value of the heat transfer coefficient on the surface of the battery α_b can be determined from the formula, which is valid for horizontal surfaces under conditions of natural convection:

$$\alpha_b = 0.59 \frac{\lambda_{air,in}}{l} (Gr_l Pr_{air})^{0.25}, \quad (2.24)$$

where $\lambda_{all,in}$—thermal conductivity coefficient of indoor air, W/m°C; Pr_{air}—Prandtl number for air; $Gr_l = \frac{g\beta\Delta T l^3}{\nu^2}$—Grasshopf number; $g = 9.8\,\text{m/s}^2$—gravitational constant; β—coefficient of volumetric expansion of air; $\Delta T = T_b - T_{air,in}$—temperature head on the surface of the battery, °C; ν—coefficient of kinematic viscosity of air, m/s^2; l—typical vertical size of the battery.

The amount of heat emitted from lighting lamps can be calculated using the formula:

$$Q_{ll} = n_l P_l \eta_l, \tag{2.25}$$

where n_l—number of lamps in the building, pcs; P_l—lamp power, W; η_l—coefficient showing how much of the power is spent on heating the external environment.

The amount of heat input from household electrical appliances can be determined from the following equation:

$$Q_{ea} = \sum\nolimits_i^m P_{ps} \eta_{ps}, \tag{2.26}$$

where m—number of devices, pcs; P_{ps}—electrical power of the source, W; η_{ps}—coefficient that shows how much of the device's power is converted into thermal energy.

The amount of heat input from the people in the room can be calculated from the ratio:

$$Q_p = n_p P_p, \tag{2.27}$$

where n_p—number of people in the building, pcs; P_p—average amount of heat released by a person into the environment.

The amount of heat that enters the room from solar radiation is determined as follows:

$$Q_{sr} = P_{sr} n_w F_w \eta_{ht} sin\gamma, \tag{2.28}$$

where P_{sr}—amount of thermal energy received from the Sun per 1 m^2 of the object normal to its surface; n_w—number of windows on which solar radiation falls, pcs; η_{ht}—coefficient of heat transmission through the window surface; γ—angle of incidence of sunlight on the window surface.

The heat losses to the environment from the building envelope are found from the expression:

$$Q_{wall} = KF_{wall}\left(T_{air,in} - T_{air,out}\right), \tag{2.29}$$

where K—coefficient of heat transfer through the surface of the building walls, W/m°C; F_{wall}—total surface area of the building walls, m^2; $T_{air,in}$, $T_{air,out}$—temperature of internal and external air, °C.

The heat transfer coefficient through a multilayer wall of a building is calculated from the following equation:

$$K = \frac{1}{1/\alpha_{in} + \sum_{i=1}^{m} \delta_i/\lambda_i + 1/\alpha_{out}}, \tag{2.30}$$

where α_{in}, α_{out}—heat transfer coefficient on the inner and outer surfaces of the wall, respectively, W/m°C; δ_i—thickness of the *i*th wall layer, m; λ_i—thermal conductivity of the material from which the *i*th layer of the wall is made.

The coefficient of heat transfer on the inner surface of the wall is calculated by the formula:

$$\alpha_{in} = C_0 \frac{\lambda_{air,in}}{L} (Gr_L Pr_{air})^n, \tag{2.31}$$

where $\lambda_{air,in}$—thermal conductivity coefficient of indoor air, W/m°C; Pr_{air}—Prandtl number for air; $Gr_L = \frac{g\beta\Delta TL^3}{\nu^2}$—Grasgoff number; $\Delta T = T_{wall,in} - T_{air,in}$—temperature head on the inner surface of the wall, °C; ν—coefficient of kinematic viscosity of air, m/s^2; L—characteristic vertical dimension of the wall.

The coefficient C_0 and the index *n*, depending on the value of the Rayleigh number $Ra = GrPr$ take the following values:

$$\text{in the range of } 10^4 \le \text{Ra} \le 10^9 \text{C0} = 0.59, \text{n} = 0.25; \tag{2.32}$$

$$\text{in the range of } Gr_x Pr \ge 10^9 C0 = 0.13, n = 0.33. \tag{2.33}$$

The outer surface of the wall may have a mixed mode of heat transfer with natural (in the absence of wind) and forced convection (in the presence of wind flowing around the building), then the heat transfer coefficient on the outer surface of the wall α_{out} can be calculated by the formula:

$$\alpha_{out} = (\alpha^3_{out,nat} + \alpha^3_{out,sed})^{1/3}, \tag{2.34}$$

where the heat transfer coefficient for natural convection $\alpha_{out,nat}$ can be calculated from equation:

$$\alpha_{out} = C_0 \frac{\lambda_{air,out}}{L} (Gr_L Pr_{air})^n, \tag{2.35}$$

where *L* is the characteristic vertical dimension of the outer wall, and the coefficient and exponent are chosen according to the range of changes in the Rayleigh number from formulas (2.32) and (2.33).

The heat transfer coefficient on the surface of the outer wall of a building is calculated from the formula:

$$\alpha_{out,in} = C_1 Re_Z{}^n Pr^m, \tag{2.36}$$

where Re_Z—Reynolds number; Z—characteristic dimensions of the outer wall in the horizontal wind direction; Pr—Prandtl number for the outer air, and the coefficient C_1 and the indicators n and m, depending on the value of the Reynolds number Re_Z take the following values:

$$\text{in the range of } Re_Z \leq 10^{10} C1 = 0.5, n = 0.25.\ m = 0.33; \tag{2.37}$$

$$\text{in the range of } Re_Z \geq 10^{10} \mathrm{C1} = 0.037, \mathrm{n} = 0.8.\ \mathrm{m} = 0.43. \tag{2.38}$$

The thermal resistance of the window Q_w according to DSTU B V.2.7-107:2008 Glued double-glazed windows for construction purposes is in the range $R = 0.32$–0.9 W/m°C, and the sunlight transmittance is in the range $\eta_{ht} = 65$–80%. Then the heat loss through the windows can be calculated using the formula:

$$Q_w = n_w K F_w \left(T_{air,in} - T_{air,out}\right), \tag{2.39}$$

where n_w—number of windows in the building; K—heat transfer coefficient through the window surface, W/m°C; F_w—total surface area of windows in the building, m^2; $T_{air,in}$, $T_{air,out}$—temperature of internal and external air, °C.

The heat transfer coefficient through the windows in a house is calculated from the ratio:

$$K = \frac{1}{1/\alpha_{in} + \sum_{i=1}^{m} \delta_i/\lambda_i + 1/\alpha_{out}}, \tag{2.40}$$

where α_{in}, α_{out}—heat transfer coefficient on the inner and outer surfaces of the windows, respectively, W/m°C; δ_i—thickness of the ith window layer, m; λ_i—thermal conductivity coefficient of the material from which the ith layer of the window is made.

To calculate the heat transfer coefficients on the surface of windows, the relations (2.31)–(2.36) are used, where the vertical and horizontal dimensions of the windows are used as a characteristic dimension in the calculation of the Rayleigh and Reynolds numbers.

The thermal resistance of translucent doors is calculated in the same way as for window doors, with a thermal resistance value in the range of $R = 0.38$–0.7. Using this data, the heat loss through the door can be calculated using the formula:

$$Q_d = n_d K F_d \left(T_{air,in} - T_{air,out}\right), \tag{2.41}$$

where n_d—number of doors in the building, K—heat transfer coefficient through the door surface, W/m°C; F_d—total surface area of the building door, m^2; $T_{air,in}$, $T_{air,out}$—is the temperature of internal and external air, °C.

The coefficient of heat transfer through a door in a house can be calculated from the following equation:

$$K = \frac{1}{1/\alpha_{in} + \sum_{i=1}^{m} \delta_i/\lambda_i + 1/\alpha_{out}}, \tag{2.42}$$

where α_{in}, α_{out}—heat transfer coefficient on the inner and outer surfaces of the door, respectively, W/m°C; δ_i—thickness of the ith layer of the door, if it has several leaves, m; λ_i—thermal conductivity coefficient of the material from which the ith layer of the door is made. When calculating the thermal resistance of a door, the data given in DSTU B B.2.7–107:2008 Glazed insulating glass units for construction purposes can be used.

To calculate the heat transfer coefficients on the door surface, the relations (2.31)–(2.36) are used, where the vertical and horizontal dimensions of the door are used as a characteristic dimension in the calculation of the Rayleigh and Reynolds numbers. Q_f—heat loss through the floor J.

The calculation of heat losses through the floor is performed by dividing the floor surface into four zones, each of which is 2 m wide (Fig. 2.5). The numbering starts with the first zone, which is close to the wall, followed by the second zone, etc. The width of the fourth zone is not limited.

The thermal resistance to heat transfer on the surface of an uninsulated floor is assumed to be as follows: for zone 1, $R_1 = 2.1$; for zone 2, $R_2 = 4.3$; respectively, for zones 3 and 4, $R_3 = 8.6$, $R_4 = 14.2$ m^2°C/W.

For such a floor, the calculation formula for determining heat loss is as follows:

$$Q_f = \sum_{1}^{4} (F_1/R_1 + F_2/R_2 + F_3/R_3 + F_4/R_4)(T_{air,in} - T_{gr}), \tag{2.43}$$

where the area of each zone is defined by the expression:

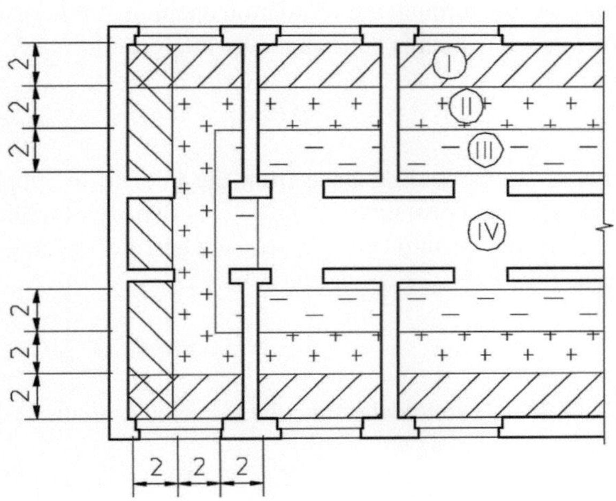

Fig. 2.5 Division of the floor into zones with different thermal resistance

$$F_1 = 2(2H + 2(L - 2));\ F_2 = 2(2(H - 2) + 2(L - 4));$$

$$F_3 = 2(2(H - 4) + 2(L - 6));\ F_4 = HL - (F_1 + F_2 + F_1 + F_2),$$

where *H, L*—length and width of the building, respectively, m; T_{gr}—soil temperature, which is selected within 8–10 °C.

If the floor has *n* layers of insulating coating, the heat loss is determined by the expression:

$$Q_f = F_f / \sum_{i=1}^{n} (\delta_i/\lambda_i)(T_{air,in} - T_{gr}), \quad (2.44)$$

where F_f—floor area, m^2; δ_i/λ_i—*is* the thermal resistance of the *i*th layer of the floor, m^2°C/W.

The calculation of heat loss through the ceiling is based on the calculation of the insulation layers used in construction. Based on this, the heat loss through the ceiling can be calculated using the formula:

$$Q_{wall} = F_{wall} / \sum_{i=1}^{n} (\delta_i/\lambda_i)(T_{air,in} - T_{air,out}), \quad (2.45)$$

where F_{wall}—ceiling area, m^2; δ_i/λ_i—thermal resistance of the *i*th layer of the ceiling, m^2°C/W. The floor area can be calculated if the dimensions of the building are known $F_{wall} = HL$, where *H* and *L*—length and width of the building.

If there are ventilation openings in the room, heat loss due to natural air ventilation is determined by the pressure difference between the room and the environment, which is caused by different temperatures of heated and cold air. To determine the air flow rate through the ventilation opening, it is necessary to determine the pressure difference, which is determined by the formula:

$$\Delta P_{air} = gh(\rho_{air,in} - \rho_{air.out}), \quad (2.46)$$

where *h*—vertical distance from the floor to the upper vent, m; $g = 9.8$ m/s^2—gravitational constant; $\rho_{air,in}$, $\rho_{air,out}$—density of internal and external air, kg/m^2.

The relationship between pressure and air flow rate in a ventilation opening $W_{нов}$ is as follows:

$$\Delta P_{air} = \rho_{air,in} W_{air}^2 / 2. \quad (2.47)$$

Now we can to determine the air velocity in the ventilation opening:

$$W_{air} = (2\Delta P_{air} / \rho_{air,in})^{1/2}. \quad (2.48)$$

The mass flow rate through the ventilation opening G_{all} will be equal:

$$G_{all} = \rho_{air,in} W_{air} F_{cs}, \tag{2.49}$$

where F_{cs}—cross-sectional area of the vent, m^2.

Thus, the total heat loss from the building due to ventilation will be determined by the expression:

$$Q_v = G_{air} c_{s,air,in} T_{air,in}, \tag{2.50}$$

where $c_{s,air,in}$—specific heat capacity of the internal air, J/kg°C; $T_{air,in}$—its temperature, °C.

Heat loss due to air permeability of the building envelope has two components—losses through windows and losses through wall structures:

$$Q_{ap} = Q_{ap,w} + Q_{ap,wall}, \tag{2.51}$$

where $Q_{ap,w}$, $Q_{ap,wall}$—losses due to air permeability of windows and walls, respectively, J.

To determine the heat loss in windows due to air permeability, it is necessary to determine the value of thermal transmittance, which is determined by the class of windows. If the class and area of the windows are known, the heat loss due to infiltration can be determined as follows.

Determine the air flow through the windows at the pressure difference, which is determined by formula (2.46):

$$G_{w,v} = n_w F_w \eta_{air,w} (\Delta P_{air}/100)/3600, \tag{2.52}$$

where $G_{w,v}$—volume air flow rate through all windows of the building, m^2; n_w—number of windows in the building, pcs; F_w—area of one window, m^2; $\eta_{air,w}$—specific air permeability for windows, which is selected depending on the window class, m^3/(m^2·h).

The amount of heat loss due to air permeability of windows is calculated by the formula:

$$Q_{ap,w} = G_{w,v} c_{s,air,in} T_{air,in}, \tag{2.53}$$

where $c_{s,air,in}$—specific heat capacity of the internal air, J/kg°C; $T_{air,in}$—is its temperature, °C.

The amount of heat loss due to air permeability through the walls depends on the number of layers of building materials in the wall and the air permeability coefficient of each layer, which are given in DSTU-N B V.2.6.-191:2013.

The total value of the thermal transmittance coefficient for a wall consisting of n layers G^{coef} can be determined by the formula:

$$G^{coef} = \left(\sum\nolimits_{i=1}^{n} \frac{1}{G_i^{\Delta P}}\right), \tag{2.54}$$

where $G_i^{\Delta P}$—air permeability coefficient for the ith layer, which is given in DSTU-N B V.2.6.-191:2013, kg/m^2.

Thus, the heat loss through the building envelope can be calculated using the formula:

$$Q_{ap,wall} = G^{coef} F_{wall} c_{s,air,in} T_{air,in} (\Delta P_{air}/100)/3600, \tag{2.55}$$

where F_{wall}—surface area of the wall envelope of the entire building without taking into account the surface area of windows and doors, m^2; $c_{s,air,in}$—specific heat capacity of the internal air, J/kg°C; $T_{air,in}$—its temperature, °C; ΔP_{air}—pressure losses, which are calculated by formula (2.41), Pa.

After calculating all components of heat losses in a building, which are determined by formulas (2.23)–(2.55), it is possible to determine the amount of heat required for heating the building Q_b

$$Q_b = Q_{wall} + Q_w + Q_d + Q_f + Q_c + Q_v + Q_{ap} - Q_{ll} - Q_{ea} - Q_p - Q_{sr}. \tag{2.56}$$

It should be noted that the calculation is made for different regions of Ukraine, as the external atmospheric conditions differ for each region.

2.3 Evaluation of the Impact of Air Permeability of Building Envelope Structures by the Method of Test Pressure and Thermal Imaging

Establishing the components of irrational actual energy consumption of the building envelope and engineering systems, determining and evaluating energy efficiency indicators, determining energy saving potential, developing a set of technical solutions to improve the energy efficiency class of the building and providing conclusions on energy saving in the building is impossible without an energy audit [8].

Depending on the purpose of the energy audit and the formulated terms of reference, the following methods are used to assess the energy performance of buildings: calculation method based on design data, calculation method based on design data and the results of technical inspections, calculation and measurement method and measurement (operational) method.

The calculation and measurement method is of the greatest interest because it is used for buildings in operation and for buildings being put into operation, to determine and conduct a detailed analysis of the actual energy and thermal performance of buildings, and to develop energy-saving measures based on the results of instrumental thermal measurements.

In this case, the energy performance of the building is determined in accordance with [4, 9] based on the results of measurements:

- actual values of air temperature in typical rooms of the building according to [10];
- actual values of the outdoor air temperature according to [10];
- actual values of the heat transfer resistance of the building envelope according to [10, 11];
- actual values of air exchange in typical building rooms and air permeability of building envelopes according to [12];
- actual heat and electricity consumption according to [13];
- moisture content of materials of the layers of the building envelope according to [10].

The standard [11] establishes a qualitative method for detecting thermal failures in the building envelope by means of thermal imaging inspection. Thermal imaging inspection in Ukraine is performed in accordance with [11]. Building thermography is a method that displays and presents the temperature distribution on the surface of the external building envelope. The measurement procedure includes an explanation of thermal images (thermograms). This method is used to identify significant deviations in the thermal properties, including the degree of airtightness, of the elements that make up the external building envelope. Two forms of thermography are presented in this standard: the infrared camera test is primarily intended to investigate the overall performance of new buildings or the result after refurbishment; the simplified infrared camera test is appropriate for external inspection, for example, on site or in the case of product control or other routine inspections.

Thermal imaging inspection of building sites involves:

- determination of the temperature distribution over the surface of the building envelope according to the distribution of the actual radiation temperature obtained using the infrared radiation sensing system;
- determining whether the surface temperature distribution is abnormal, i.e. caused by, for example, insulation defects, moisture content and/or air filtration;
- during confirming the condition of the previous paragraph, it is necessary to determine the type and extent of structural defects.

To determine the abnormality of the thermal insulation properties based on the observed deviations in the obtained thermograms, they are compared with the expected temperature distribution over the surface, determined by the designed characteristics of the building envelope under the environmental conditions observed during the inspection. The expected temperature distribution can be determined by means of "reference" thermograms, calculations or other studies [11]. This determination is based on drawings and other documents that characterise the external envelope and heating and ventilation systems of the inspected facility.

The following factors are taken into account to determine the measurement programme, especially the choice of the surface of the enclosing structure (external or internal) to be inspected by thermal imaging:

(a) technical parameters and characteristics of thermal imaging equipment;
(b) characteristics of the building envelope, relevant types, location of heating systems, structural elements and insulation layers;

(c) radiative properties of the surface, for example, the finishing material;
(d) climatic conditions;
(e) accessibility for easy inspection;
(f) environmental impact;
(g) other important factors.

The difference in temperature between the two sides of the building envelope should be sufficient to detect thermal failures. For ease of interpretation, it is best to carry out a thermal imaging inspection when the temperature and pressure differences on both sides of the building envelope are constant. This implies, among other things, that the test should not be carried out when the external or internal air temperature is subject to significant changes, or when the building is exposed to direct sunlight, or when the wind changes significantly. The actual requirements may vary according to the thermal properties of the building structure under test and the characteristics of the infrared sensing system used and the local climate conditions.

If air infiltration can affect the results of the inspection, it is necessary to create a pressure difference on both sides of the building envelope or to conduct the inspection under conditions where there is a pressure difference. If the main purpose of the thermal imaging inspection is to determine air infiltration, then the pressure difference should be at least 5 Pa at the inspection site. Thermal imaging inspection is performed from the lower pressure side.

When inspecting the building envelope, it is necessary to determine the influence of ventilated air spaces, e.g. in walls or heat sources (if any) installed in the building (built-in pipes, chimneys, etc.) that affect the temperature of the structures under investigation. Heat sources that affect the results of the survey must be disconnected before the survey begins. Furniture, paintings and other items that may affect the results should be removed to make the areas to be inspected clear. Necessary changes should be carried out in such a way as to avoid their temporary effects.

Immediately prior to the survey, the temperature of the internal and external air should be determined with an absolute error of ± 1 K. When the pressure difference on both sides of the structure is to be determined, it is recommended to measure it with an error of ± 2 Pa on the leeward and windward sides of each floor. It is particularly important to determine the direction of the pressure difference through the building envelope and the location of the neutral plane, if any.

Changes in the actual radiation temperature along the thermal image of the building envelope surface should be measured with an error of $\pm 10\%$ or ± 0.5 K, whichever is greater. When measuring the temperature of the "reference" surface, it is recommended to determine it with an error of ± 0.5 K.

The inspection should begin with a preliminary test of the structure's surface. Surface areas of particular interest or with obvious abnormalities should be examined in detail. Thermograms should be obtained from areas of the structure under investigation that are free of defects and areas where structural defects are possible. If the thermograms show air filtration, this should be confirmed by air velocity measurements, if possible. If necessary, the factors affecting the heat transfer coefficient of the internal surface of the structure (air filtration, heat radiation, moisture condensation) should be assessed.

Three types of defects are usually detected by thermal imaging:

- temperature (thermal);
- humid;
- infiltration.

The first two types of defects require an appropriate temperature head, which should be at least 15°. The third type requires a steady pressure drop, which is not always present, especially in the off-season.

When identifying air infiltration locations, the operator is faced with uncertainty in the distribution of pressure differences between the interior and exterior of the building structure. This uncertainty arises from the following factors:

- temperature difference between indoor and outdoor air;
- height of the building;
- the impact of wind on the structure;
- the impact of mechanical ventilation.

The above factors and their interaction can affect the direction of infiltration flows or lead to their absence. Therefore, the defect may not be detected if the air pressure drop vector, i.e. the air flow at the time of the survey, is directed away from the operator.

A high-quality result can be obtained if the wind is directed at the part of the façade under study during the survey or if there is excessive pressure or vacuum in the building compared to the ambient pressure. This can be done with the help of special equipment for determining the air permeability of buildings, which is called a Blower Door (Fig. 2.6). This device consists of a fan or blower attachment to a door or window that is adjustable to fit the normal door or window openings. The fan or blower should have a variable speed motor to provide the range of airflow rates required [12].

It should be noted that this system can be either single or multi-fan, which is used to determine the air permeability of large buildings (Fig. 2.7). In some cases, the fans of the building's ventilation system can be used to create increased or decreased pressure. A preliminary site survey is recommended to determine the number of supply or exhaust fans with the appropriate air capacity and to determine whether fans operating with 100% outside air or 100% exhaust air can be used [12]. The available means of controlling the supply (or exhaust) air flow rate (e.g., using adjustable air dampers or controlling the fan impeller speed) should be inspected and evaluated. You can also check the ductwork system and select appropriate locations for the airflow measurement device.

Thermal imaging inspection, carried out under guaranteed pressure drop, allows:

- eliminate the influence of factors that reduce the probability of defect detection (e.g. wind, chimney effect, etc.);

Fig. 2.6 Layout and appearance of the air gate

- identify filtration locations at low temperature differences (for example, in summer or at facilities where this difference is insignificant);
- locate filtration points when measuring the air permeability of the envelope.

The current version of [14, 15] stipulates that a prerequisite for designing new construction, reconstruction, overhaul, and thermal modernisation is to achieve an energy efficiency class not lower than class C. One of the prerequisites for achieving this class is that the building envelope complies with the air permeability standards defined in [16]. All buildings of energy class C, B or A and their parts shall be designed and constructed so that their air exchange measured in accordance with [12] at a pressure difference between the inside and outside of 50 Pa does not exceed

Fig. 2.7 Multi-fan system for testing buildings for air permeability

the values of the air exchange rate for residential, public, administrative, educational and medical buildings defined in [17, 18].

The essence of the method is that an air flow is passed through the object under test and, after the steady-state flow is established, the air flow rate and pressure drop between the volume and the environment or between opposite surfaces of the structure are measured. Tests are carried out at lower pressure and at higher pressure relative to the environment. The measurement results are used to calculate the generalised air permeability characteristics of the test object or the air permeability resistance of the envelope (the lower of the two results is taken) (Fig. 2.8) [12].

The essence of the system is to create a vacuum or overpressure of 50 Pa inside the building, which is equivalent to the simultaneous wind pressure on the building envelope at a speed of 9–10 m/s. It is impossible to achieve such conditions naturally, so this equipment is indispensable when conducting a thermal imaging survey to identify places of air infiltration and exfiltration.

The test shall be carried out by measuring the air volume flow rate and the internal and external pressure drop over the range of the generated pressure drop increase at intervals not exceeding 10 Pa. At each stage of the pressure increase, at least five measurements shall be made at evenly spaced intervals between the lowest and highest pressure drops. The lowest differential pressure shall be approximately 10 Pa ($\pm$3 Pa tolerance) or five times the value of the pressure drop at zero flow ($\Delta p01$), whichever is greater. The highest differential pressure should be at least 50 Pa, but for maximum accuracy of the calculation results, it is recommended to measure at a pressure drop of up to 100 Pa [12].

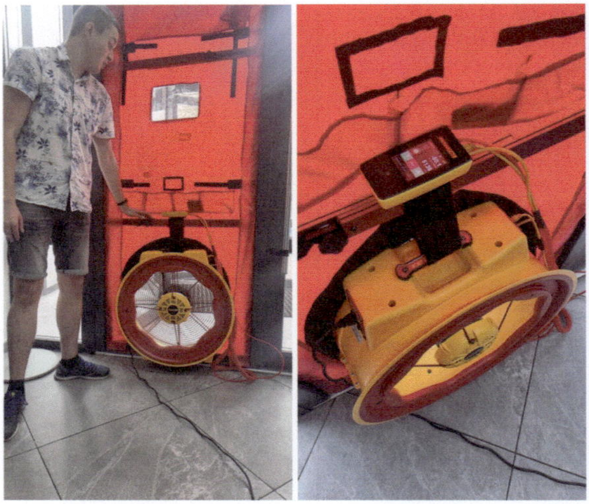

Fig. 2.8 Operation of the blower door system

However, due to the large size of many public buildings and the limited capacity of the mobile air pressure equipment used for the test, a pressure difference of 50 Pa may not be achievable. In these cases, either additional air pressure equipment should be used, or equipment with a higher capacity (to increase the overall capacity), or the test should be performed to the highest pressure differential that can be produced by the available equipment. In general, the test is considered invalid if a pressure drop of 25 Pa cannot be achieved. Cases where the highest differential pressure is between 25 and 50 Pa shall be clearly recorded in the test report, indicating that the requirements of this standard have not been fully met and explaining the reasons. Alternatively, it is possible to test large buildings by dividing them into several separate parts.

Several test methods are considered in [12], depending on the purpose of the test.

(a) Method 1—testing the building under operating conditions when the natural ventilation openings are closed and the mechanical ventilation or air conditioning openings of the entire building are sealed.
(b) Method 2—testing of a separate area of the building envelope when all openings for natural ventilation are hermetically sealed and doors, windows and hatches are closed.
(c) Method 3 is a task-specific building test where openings are sealed or opened in accordance with the task, which is developed on the basis of relevant national regulations or standards.

The volume of a building or part of a building to be tested depends on the purpose of the test and is defined as follows [12]:

(a) the part of the building prepared for the measurement shall include all ventilated and air-conditioned rooms (i.e. rooms with direct or indirect heating, cooling and/or mechanical ventilation);

(b) if the purpose of the measurement is to assess the fulfilment of the airtightness requirements in accordance with the applicable building regulations or standards that define the volume of the building to be tested;
(c) if the purpose of the measurement is to assess the fulfilment of airtightness requirements in accordance with applicable building regulations or standards and the building volume to be tested and measured is not defined in these building regulations or standards, then it shall be determined as in point (a);
(d) in special cases, the volume of the building for testing and measurement can be determined by agreement with the customer.

Each individual part of the building can be measured individually, for example, each apartment in a multi-apartment building. However, when processing the results, it should be borne in mind that the air flows determined in this way may include air flows through leaky areas in adjacent parts of the building.

Table 2.1 shows the classification of the holes and their state during the measurement, depending on the method used.

Detecting the flow location is the next important step and is required to reduce the flow area and to assess the flow distribution of buildings. According to [12], the following methods are used.

1. Subtraction method. The areas of the building envelope and/or structural elements to be analysed for leakage are covered with a protective film. After determining the flow area with and without the protective film, the difference between the two provides the flow area.

Table 2.1 Condition of the holes during the measurement

Classification of openings in buildings	Method 1	Method 2	Method 3
	The building in service conditions	Separate area of enclosing structures	Task-specific testing
Ventilation openings for natural ventilation	Closed	Hermetically sealed	Closed, hermetically sealed or open as specified
Openings for mechanical ventilation or air conditioning of the entire building	Hermetically sealed	Hermetically sealed	Closed, hermetically sealed or open as specified
Openings for mechanical ventilation or air conditioning (intermittent use only)	Closed	Hermetically sealed	Closed, hermetically sealed or open as specified
Windows, doors and hatches in the building envelope	Closed	Closed	Closed, hermetically sealed or open as specified
Openings not intended for ventilation	Closed	Hermetically sealed	Closed, hermetically sealed or open as specified

2. Use of infrared thermal imaging. During a reduced-pressure test, a thermal imaging camera can be used to locate air infiltration if there is a difference between the building's internal and external temperatures.
3. Use of smoke. Smoke is generated to visualise airflow through an enclosure, device, etc. and to identify the location of the flow. This method may require practical skills, such as smoke generation rates. Alternatively, it is possible to use the hands to feel the airflow around devices, etc. on the envelope, but this method is not accurate due to individual characteristics.
4. Use of an anemometer. During the pressurisation or depressurisation test, an anemometer can be placed at points on the enclosure or device where air flow is possible. If the device indicates the presence of air flow, this indicates flow.

Taking into account the previously taken into account calculation and measurement method used for buildings in operation and for buildings under commissioning to determine and conduct a detailed analysis of the actual energy and thermal performance of buildings that affect the energy consumption for heating, cooling, hot water supply and lighting of the building and the efficiency of energy use [4, 6, 19, 20] are assessed on the basis of data obtained during the energy audit, which is the basis for drawing up an energy balance.

In order to understand the difference between the survey with and without a guaranteed differential pressure, the following thermograms are presented, which were obtained during field tests of residential, public and industrial buildings using the method defined in [12]. The repository data [25] contains examples of reports based on tests of air permeability of building envelopes using the test pressure method [12].

After conducting a "Blow Door Test" of the residential building, the following conclusions were drawn:

1. Thermal imaging survey of the building facade (external walls) did not reveal any thermal failures, the temperature distribution is uniform.
2. Thermal imaging of the cold attic ceiling (image IR000058.BMT) revealed a "cold bridge" at the junction of the outer walls and the ceiling (kitchen-living room). The probable cause is a loose connection between the floor insulation (mineral wool) and the walls.
3. During the building leakage test, the following locations were identified:
   - junction of the window sill to the window frame—due to the absence of steam and water barrier tapes during window installation (requirements of DSTU B V.2.6.79–2009 "Joints of window junctions");
   - Attic ceiling adjoining the exterior walls (entrance hall, image IR000053.BMT)—the probable cause is leaky gluing of vapour barrier films to the exterior walls.
   - holes in the drywall for electrical communications.
4. The external entrance door must be adjusted for a snug fit.

Based on the results of the research of the residential building, the following recommendations were made:

1. Eliminate leaks at the junction of the window sill to the window frame. Apply silicone sealant to the junction between the window and the sill from the inside of the house. Additionally, after the heating season, it is recommended to dismantle the window sill and check the sealing condition of the window-wall junction. If there are any holes, additionally treat the installation seams with a one-component acrylic vapour-permeable sealant (e.g. STIZ-A), which will protect the installation foam from moisture penetration and weathering, in particular sunlight. STIZ-A has sufficient vapour permeability to remove moisture from the assembly and insulation.
2. Sealing the junction of the attic ceiling to the outer walls (hallway) with mortar or sealant, installing a decorative skirting board.
3. General recommendations for the construction of such houses:
   - It is recommended to insulate the gas block or foam block with a more vapour-permeable insulation (e.g. mineral wool) to ensure sufficient moisture removal through the walls. Expanded polystyrene is more than 10 times less vapour-permeable than mineral wool, which creates a risk of moisture accumulation in gas block walls.
   - The walls should be insulated with at least 150 mm of mineral wool to meet the minimum requirements for heat transfer resistance (requirements of DBN B 2.6-31-2021 "Thermal insulation and energy efficiency of buildings"), which will help reduce the specific energy consumption of primary energy sources [21].
   - Seal the joints of the vapour barrier with each other and with the walls.
   - When installing windows and doors, use steam and hydrobarrier tapes, the thermal characteristics of which are given in [22, 2324].

Install metal insulated doors with a triple layer of seals.

After conducting a "Blow Door Test" of the office space, the following conclusions were drawn:

- the main heat losses are due to leaky ceiling and suspended ceiling structures (SR000520.BMT, SR000524.BMT);
- thermal imaging detected heat loss due to loose fitting of the door and installation (SR000527.BMT);
- The tightness of the installation of window structures was checked (SR000528.BMT, SR000529.BMT, SR000530.BMT);
- significant heat loss through the ventilation and air conditioning system (IR000532.BMT).

After conducting a "Blow Door Test" of the warehouse building, the following conclusions were drawn:

- the main air heat loss occurs through the joints of the wall/roof/floor bearing structures (SR000488.BMT, IR000489.BMT, SR000519.BMT, IR000514.BMT, IR000511.BMT, IR000510.BMT, IR000512.BMT);

– thermal imaging detected areas of increased heat loss due to loose fitting of door structures to the floor (IR000509.BMT).

The Blow Door Test allows to quickly and accurately assess the energy efficiency of a room.

2.4 Conclusions

The paper proposes a model of the thermal regime of a building as a single heat and energy system, taking into account its three subsystems: a model of the heat and energy impact of the external climate on the building (η_1), a model of the heat and energy impact of the thermal inertia of the building envelope (η_2) and a model of the heat and energy impact of the building's engineering systems $(\eta)_{0.3}$.

To evaluate the specified indicators, the project deviation coefficient (PD) is introduced, which shows (in fractions of a unit) the deviation of these indicators from the design conditions and characterises the degree of difference between the implemented solution and the optimal one and is an indicator of the thermal energy efficiency of the project solution as a whole.

The paper estimates heat energy savings when using different variants of automatic balancing—for a single-pipe heating system with a constant hydraulic mode and separately for a single-pipe heating system with a variable hydraulic mode.

The article presents an improved mathematical model for calculating heat losses in buildings, which additionally takes into account heat losses due to air permeability of building envelopes, as well as refined methods for calculating convective flows on wall surfaces and ventilation losses.

The necessity of conducting a thermal imaging inspection of buildings is substantiated, the types of defects that can be detected during such an audit and the conditions under which the inspection will allow to identify all defects are presented. It has been established that the uneven thermal properties of the elements of the external building envelope lead to changes in the temperature on its surface. The nature of the surface temperature distribution can be used to detect thermal failures due to, for example, defects in insulation, moisture content and/or air infiltration in the elements of the building envelope.

Field studies of buildings for various purposes have been carried out, which indicate the need to take into account the pressure difference between the outside and inside of the building. It was found that the surface temperature is also affected by the air flow in the building and/or through its envelope. The analysis of thermograms showed that there is an obvious difference in the quality of defect detection without and with the additional creation of a pressure difference, which can only be provided by specialised equipment for determining the air permeability of buildings "Blower Door".

References

1. DSTU-N B V.1.1-27: Building climatology. Kyiv: Ministry of the Region of Ukraine, 2011, 130p. (2010)
2. Bolharova, N.M., Ploskyi, V.O., Skochko, V.I.: Practical aspects of constructing a physical discrete heat exchange model of an energy efficient building. Technol. Eng. Des. Kyiv, KNUBA **13**, 9–20 (2018). https://doi.org/10.32347/2310-0516.2018.11.7-21
3. DBN B.2.6-31: Thermal insulation and energy efficiency of buildings. State Standard of Ukraine, Kyiv, 2022, 33c. (2021)
4. DSTU 9190: Energy efficiency of buildings. Method for calculating energy consumption during heating, cooling, ventilation, lighting and hot water supply (2022)
5. ISO 52016-1: Energy performance of buildings - Energy needs for heating and cooling, internal temperatures and sensible and latent heat loads - part 1: calculation procedures (2017)
6. DSTU B EN ISO 13790: Energy performance of buildings. Calculation of energy consumption for heating and cooling (EN ISO 13790:2008, IDT) (2011)
7. EN ISO 13790: Energy performance of buildings - calculation of energy use for space heating and cooling (2008)
8. DSTU B B.2.2-39: Methods and stages of energy audit of buildings (The standard was developed taking into account the requirements of DSTU 4065, DSTU EN 16247-2.) (2016)
9. DSTU-N B A.2.2-13: Energy efficiency of buildings. Guidelines for the energy assessment of buildings (2015)
10. DSTU-N B V.2.6-101: Structures of buildings and structures. Method for determining the heat transfer resistance of building envelopes (2010)
11. DSTU B EN 13187: Thermal performance of buildings. Qualitative detection of thermal failures in the building envelope. Infrared method (EN 13187:1998, IDT) (2011)
12. DSTU EN ISO 9972: (EN ISO 9972:2015, IDT; ISO 9972:2015, IDT) "Thermal performance of buildings. Determination of air permeability of buildings. Test pressure method", adopted by the translation method, is identical to EN ISO 9972:2015 (en) "Thermal performance of buildings - Determination of air permeability of buildings - Fan pressurisation method (ISO 9972:2015)" (2022)
13. DSTU B V.2.2-21: Buildings and structures. Method for determination of specific heat consumption for heating of buildings (2008)
14. DBN B 2.6-31: Thermal insulation and energy efficiency of buildings (2021)
15. Order of the Ministry of Communities and Territories Development of Ukraine dated 27 October 2020 No. 260 "On Approval of Minimum Requirements for Energy Efficiency of Buildings", registered with the Ministry of Justice of Ukraine on 18 December 2020 under No. 1257/35540 (2020)
16. DSTU-N B V.2.6-191: Guidelines for the calculation of air permeability of enclosing structures (2013)
17. DSTU B EN 15251: Calculated indoor climate parameters for the design and assessment of energy performance of buildings in relation to air quality, thermal comfort, lighting and acoustics (EN 15251:2007, IDT) (2011)
18. DBN B.2.5-67: Heating, ventilation and air conditioning (2013)
19. DSTU B EN 15459: Energy performance of buildings. Procedure for the energy assessment of building systems (EN 15459:2007, IDT) (2014)
20. DSTU B EN 15603: Energy performance of buildings. Total energy consumption and energy assessment (EN 15603:2007, IDT) (2013)
21. Order of the Ministry of Communities and Territories Development of Ukraine dated 27 October 2020 No. 261 "On Approval of Amendments to the Methodology for Determining the Energy Efficiency of Buildings", registered with the Ministry of Justice of Ukraine on 17 December 2020 under No. 1254/35537 (2020)
22. DSTU B V.2.7-182: Building materials Methods for determining the effective service life and thermal conductivity of building insulation materials under design and standard conditions (2009)

23. DSTU 9191: Thermal insulation of buildings. Methods of selection of thermal insulation material for building insulation (2022)
24. DSTU B EN 15217: Energy performance of buildings. Methods for the presentation of energy performance and energy certification of buildings (EN 15217:2007, IDT) (2013)
25. Reports based on tests of air permeability. https://doi.org/10.13140/RG.2.2.22081.47200

Chapter 3
Modelling of Air Exchange in Premises

3.1 Parameters of Simulation Modelling of Indoor Air Exchange

The methodological basis of the research was formed by the following: methods of mathematical simulation modelling, a systematic approach taking into account temperature, weather and operational factors.

Usually, in mathematical models for determining the energy demand of buildings, the ventilation component (air exchange) is specified through the value of the air exchange rate. Air exchange is difficult to determine experimentally. Even with the same window design, in terms of air permeability, different amounts of air naturally enter the room. Air exchange in a room depends on a number of factors, both external and internal. The paper uses a generalisation of the methods for determining the air exchange rate based on the pressure difference given in studies [1–3] and ASHRAE approaches [4, 5], which allows the use of the above methodology for multi-storey buildings. The pressure difference that determines the air exchange in a building is created by three different mechanisms: the stack effect, wind pressure, and forced pressure of the mechanical ventilation fan [3, 6]:

$$\Delta P_{tot} = \Delta P_s + \Delta P_w + \Delta P_v = \Delta P_{\text{inf}} + \Delta P_v, \tag{3.1}$$

where ΔP_{tot}—total pressure difference, Pa; ΔP_s—pressure difference from the stack effect, Pa; ΔP_w—pressure difference from wind, Pa; ΔP_v—pressure difference from forced ventilation, Pa; ΔP_{inf}—infiltration pressure difference, Pa (Fig. 3.1).

The stack effect is also known as the buoyancy effect, which is created by the density difference between warm and cold air. Decrease in air pressure with height, determined by formula (3.2):

$$\Delta P_s = P_e(z) - P_i(z) = z(\rho_e - \rho_i)g, \tag{3.2}$$

A. Zaporozhets et al., *Buildings Energy Audit*, Studies in Systems, Decision and Control 641, https://doi.org/10.1007/978-3-032-07200-9_3

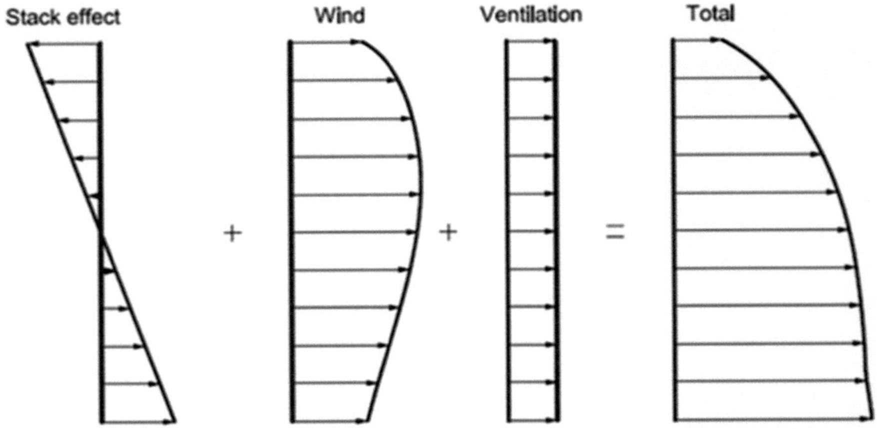

Fig. 3.1 Example of building height distribution and summation of pressure difference profiles

where Z—height from the reference point, m; ρ_e, ρ_i—air density of the external and internal air, kg/m^3; g is the free fall acceleration, m/s^2.

From the level of neutral pressure towards the first floor, the pressure difference is positive, and towards the last floor it is negative. Assuming that air is an ideal gas, formula (3.2) takes the form:

$$\Delta P_s = 3456 \cdot z\left(\frac{1}{T_e} - \frac{1}{T_i}\right), \tag{3.3}$$

where T_e, T_i are the temperature of the outside and inside air, respectively, K.

In Fig. 3.2, the height represents the neutral pressure level (NPL), at which the internal and external pressures are equal, usually found on the middle floor of a building.

Wind pressure is created when a stream of air hits an obstacle. The amount of wind pressure depends on the wind speed and direction (windward, leeward, etc.) [3].

The following Eq. (3.4) is used in most software products for modelling wind pressure [1–3]:

$$\Delta P_w = \frac{\rho U_{met}^2}{2} C_h C_p(\theta), \tag{3.4}$$

where ΔP_w is the wind pressure drop, Pa; ρ is the ambient density, kg/m^3; U_{met} is the wind speed according to the nearest weather station, m/s; C_p (θ) is the wind pressure coefficient that takes into account the wind direction through the angle of approach θ; θ is the value of the wind angle relative to the normal to the surface under consideration, °; C_h is the wind pressure coefficient that takes into account the number of storeys.

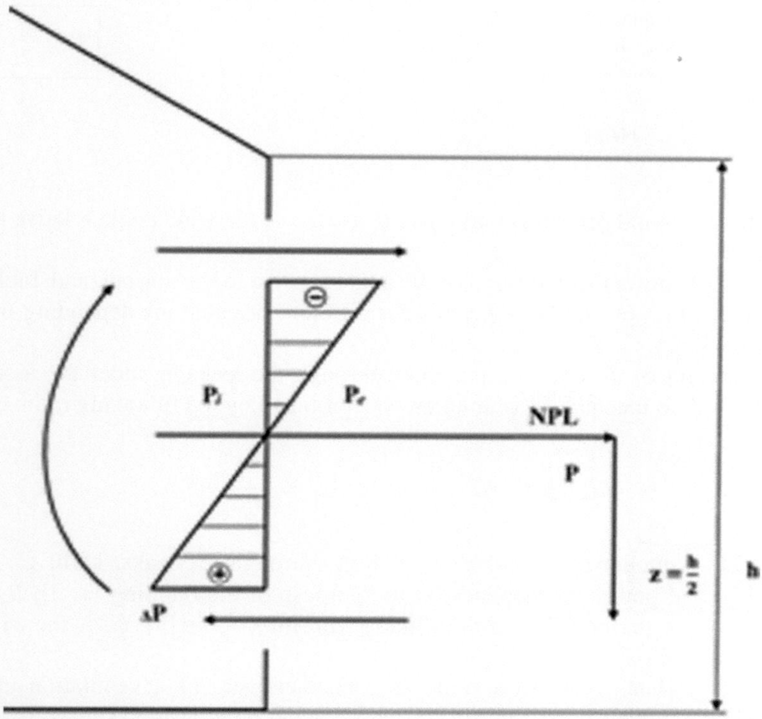

Fig. 3.2 Pressure difference from the neutral pressure level

$$C_h = \frac{U_H^2}{U_{met}^2} = A_0^2 \left(\frac{H}{H_{ref}} \right)^{2a}, \tag{3.5}$$

where U_H is the wind speed at the highest point of the object, m/s; A_0 is the coefficient that takes into account the wind shelter for a particular area; [2, 3] H is the height of the considered floor of the building from ground level, m; H_{ref} is the height at which the weather station measures the wind speed, m; *a is* the exponential coefficient that takes into account the wind shelter for a particular area [7].

The peculiarity of the wind direction is taken into account by the coefficient C_p. The wind pressure coefficient for a certain angle is determined by the formula [8]:

$$C_p(\theta) = \frac{1}{2} \left\{ \begin{array}{l} \left[C_p(0^\circ) + C_p(180^\circ)\right] \cdot \left(\cos^2(\theta)\right)^{\frac{1}{4}} \\ + \left[C_p(0^\circ) - C_p(180^\circ)\right] \cdot \left(\cos^2(\theta)\right) \\ + \left[C_p(90^\circ) + C_p(270^\circ)\right] \cdot \left(\sin^2(\theta)\right)^2 \\ + \left[C_p(90^\circ) + C_p(270^\circ)\right] \cdot \left(\sin^2(\theta)\right) \end{array} \right\}, \tag{3.6}$$

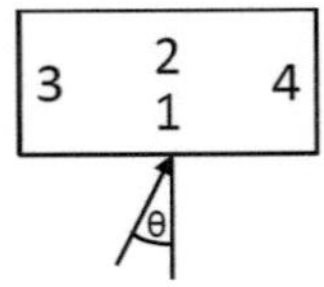

Fig. 3.3 Determination of the angle θ relative to the normal (1, 2, 3, 4—surface designations)

where $C_p(\theta)$—wind pressure coefficient, θ—value of the wind angle relative to the normal, °.

Figure 3.3 shows the definition of the angle relative to the normal, and Table 3.1 shows the values of wind coefficients for different orientations depending on the wind direction.

The amount of air entering the room through the leakage under the assumed conditions of no mechanical ventilation is determined by the following ratio:

$$G_{inf} = C\left(\Delta P_{inf}\right)^p \quad or \quad G_{inf} = \frac{\Delta P_{inf}^p}{R_{a\,\cot p}} F_w, \tag{3.7}$$

where G_{inf}—amount of air entering the room through the leaks, kg/h; C_p—this coefficient and power law depends on the purpose of the building [1, 3]; $R_{a.p.}$ —resistance to air permeability of the window, $(m^2 \cdot h \cdot Pa^{2/3})$/kg [9]; F_w is the window area, $m^2 \cdot c$.

The air exchange rate in a room as a characteristic of a ventilation unit in mathematical models is determined by the following relationship [3]:

$$n = \frac{G_{\text{inf}}}{\rho V}, \tag{3.8}$$

where V—volume of the room, m^3; n—air exchange rate, h^{-1}.

This methodology allows to take into account the flows of infiltration and exfiltration of air into the room zone, which significantly affects the energy characteristics.

Assumption: the pressure inside all rooms is the same.

Table 3.1 Calculated values of the wind pressure coefficient $C_p(\theta)$ for all enclosures

Location	Wind angle of approach to the surface relative to the normal (θ)							
	0°	45°	90°	135°	180°	225°	270°	315°
Surface 1	0.25	0.06	−0.35	−0.6	−0.5	−0.6	−0.35	0.06
Surface 2	−0.5	−0.6	−0.35	0.06	0.25	0.06	−0.35	−0.6
Surface 3	−0.6	0.2	0.4	0.2	−0.6	−0.6	−0.3	−0.6
Surface 4	−0.6	−0.6	−0.3	−0.6	−0.6	0.2	0.4	0.2

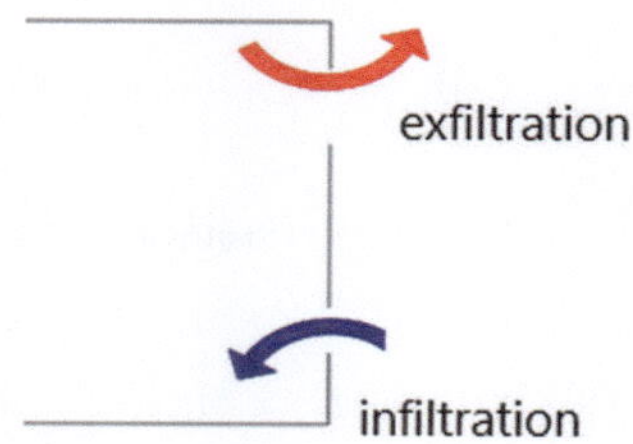

Fig. 3.4 Air exchange mechanism in a multi-storey building

This methodology allows taking into account the flows of air infiltration and exfiltration from/into the room area (Fig. 3.4), which significantly affects the energy performance.

3.2 Model of the Relationship Between CO_2 Concentration and Air Exchange

Based on the balances of air flows and CO concentrations$_2$ in the room (Fig. 3.5), the following relationships were obtained [10].

The concentration of CO_2 in the room is determined:

$$C_i = \frac{m}{V_r}. \tag{3.9}$$

The change in the amount of CO_2 over time is given by the equation:

$$\frac{dM}{d\tau} = m + V \cdot C_e - \frac{V}{V_r} M. \tag{3.10}$$

The amount of CO_2 after a time interval Δτj:

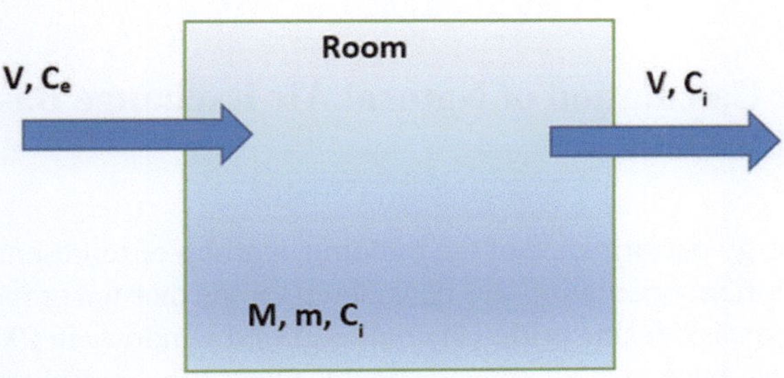

Fig. 3.5 Scheme of air exchange and changes in CO_2 concentration in a representative room: V_r—room volume, m^3; V—air exchange, m /h; ^{3} m—CO_2 emission in the room, g/h; M—mass of CO_2 in the room, g; M_0—initial mass of CO_2 in the room, g C_e, C_i, C_{i0}—concentration of CO_2 in the supply air, in the room at a given time and initially, g/m^3

$$M_j = M_{j-1} + \underline{m_j} \vartriangle \tau_j + \underline{V_j} \cdot \underline{C_{e.j}} \vartriangle \tau_j - \frac{\underline{\underline{V_j}}}{V_r} \cdot \frac{M_{j-1} + M_j}{2} \vartriangle \tau_j. \quad (3.11)$$

After n time intervals $\Delta\tau_j$, formula (3.11) takes the form:

$$M_n = M_0 + \sum_{j=1}^{n} m_j \vartriangle \tau_j + \sum_{j=1}^{n} V_j \cdot C_{e.j} \vartriangle \tau_j - \sum_{j=1}^{n} \frac{V}{V_r} \cdot M_j. \quad (3.12)$$

Equation (3.12) can be used to determine the mass of CO_2 in the indoor air through the values of C_e, V and m for each time interval $\Delta\tau_j$.

Integration of Eq. (3.10) at constant values of C_e, V and m gives an expression for the mass of CO_2 in the indoor air at any time τ:

$$M = \frac{V_r}{V}\left\{m + V \cdot C_e + \left[\frac{V}{V_r} \cdot M_0 - (m + V \cdot C_e)\right] \cdot e^{-\frac{V}{V_r}\tau}\right\}. \quad (3.13)$$

From formula (3.13), it follows that the internal concentration of CO_2 at constant values of C_e, V and m is determined:

$$C_i = \frac{m}{V} + C_e + \left[C_{i0} - \left(\frac{m}{V} + C_e\right)\right] \cdot e^{-\frac{V}{V_r}\tau}. \quad (3.14)$$

For the limiting cases, formula (1.7) takes the form (1.8)–(1.9):

$$\text{if } m \to 0,\ C_i = C_e\left(1 - e^{-\frac{V}{V_r}\tau}\right) + C_{i0} \cdot e^{-\frac{V}{V_r}}; \quad (3.15)$$

$$\text{if } m \to 0 \quad i \quad C_i >> C_e,\ C_i = C_{i0} \cdot e^{-\frac{V}{V_r}\tau}. \quad (3.16)$$

The experimental data on changes in the CO_2 concentration in the room over time were processed by Eq. (3.14).

3.3 Hourly Calculation of Natural Air Exchange Based on IWEC Data

To study the energy performance of the building, a group of representative rooms of southern and northern orientation was considered for the thermal properties of enclosures typical of mass housing with: (1) double-glazed windows in PVC profiles; (2) triple glazing in wooden twin glazing; (3) double glazing in wooden twin glazing (old translucent structures). The building has 8 floors, where a constant indoor temperature of 18 °C is maintained. Based on the selected representative rooms (on the 1st, 3rd, 6th floor of the northern and southern orientation), a study of changes in the air exchange rate in a multi-storey building with different types of air permeability of windows was carried out using the above methodology. The hourly values from the IWEC file for the climatic conditions of Kyiv [11] were used: dry thermometer temperature, wind speed and direction, and barometric pressure. The data acquisition procedure was based on the selection of a typical year within an 18-year weather data sequence [12].

The IWEC typical year climate file [11] contains hourly wind speed and direction data. The wind direction is given in degrees. The wind direction in degrees is given clockwise from north (0 °C).

The IWEC analysis of a typical year file for the city of Kyiv [11] shows that the most typical wind direction for the cold season is north and north-west, with an average speed of 2.7 m/s, which is similar to the data of the Building Climatology of Ukraine for this region [13]. The range of fluctuations in hourly wind speeds is from 0 to 18 m/s (the standing time is insignificant). The wind direction changes on average every two hours. According to the IWEC, the average outside temperature for the heating season is 1.7 °C, which is slightly different from the current climatology [13] —minus 0.1 °C. The minimum outdoor temperature in the IWEC file is −15.8 °C in February. Inter-day values of the average daily outdoor temperature can vary up to 4 °C, and fluctuations within a day can be around 5 °C [8].

Figure 3.6 shows a graph of the pressure change profiles along the height of the stack effect, depending on changes in ambient temperature.

This phenomenon is associated with the difference in density between warm and cold air and does not depend on the orientation of the room. Provided that the outside air temperature is close to the inside (during peak solar activity in the off-season), the pressure difference from the stack effect is levelled.

The study investigated the variation of pressure differences with height for different wind speeds and directions and their impact on leeward and upwind enclosures. Figure 3.7 shows the corresponding graphs of the wind pressure profiles on the south side of the building, depending on the building floor and wind direction.

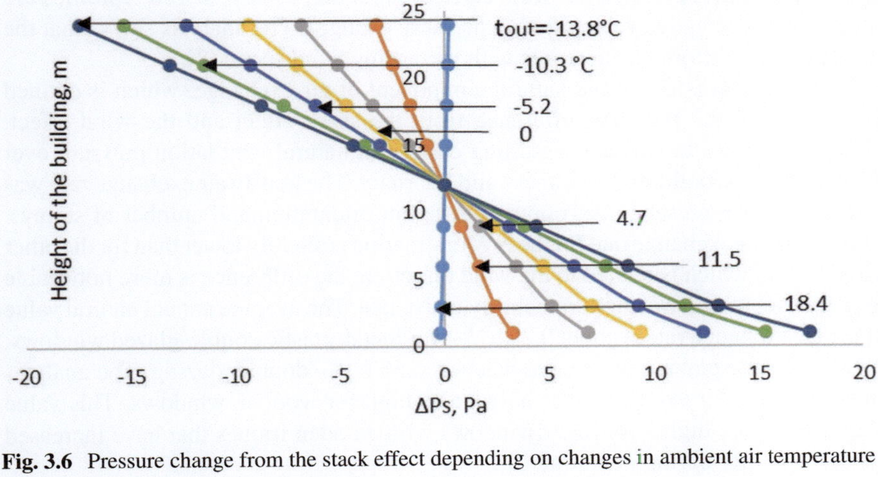

Fig. 3.6 Pressure change from the stack effect depending on changes in ambient air temperature

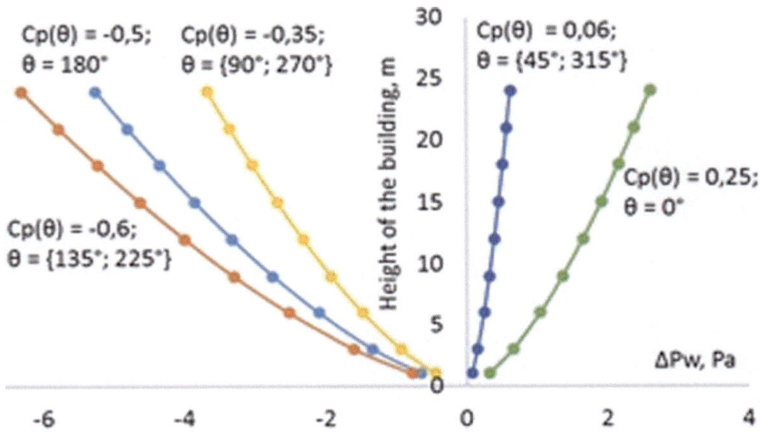

Fig. 3.7 Change of wind pressure depending on the angle of approach of the wind direction θ

The representative days were chosen when the wind speed was 3 m/s (typical wind speed for Ukrainian conditions [13]) and the outside temperature was 3 °C (average typical outside temperature for December in Kyiv [14]). The graphs show that the magnitude of the pressure change due to wind increases with height. On the windward side of the building, the wind pressure is positive, and on the leeward side it is negative [8].

Provided that the wind direction vector is directed to the surface in the frontal direction, it causes positive values of wind pressure changes, while for the leeward sides it is negative. As mentioned above, the city of Kyiv is characterised by a north and north-west wind direction. The average wind speed is 2.7 m/s, with wind speeds of more than 8 m/s typical for the north-west direction. The wind pressure change was studied for different profiles of wind speed change for the wind angle $\theta = 0°$ (Fig. 3.8). Similar calculations were carried out at angles of $\theta = 180°$ (profiles are similar with a negative value of wind pressure change). The analysis shows that the wind pressure difference increases with increasing wind speed [8].

The study investigates the natural component of air exchange, which is defined as the sum of the pressure difference from the stack effect and the wind effect. Figure 3.9 shows the average resulting change in natural ventilation pressure over the height of the building for January and February. The hourly air exchange rate was calculated for representative rooms of different orientation and number of storeys. The natural air exchange rate for the NW orientation is slightly lower than for the other orientations, which is related to the wind direction, the difference is more noticeable for windows with lower air permeability resistance. The average annual natural value of the air exchange rate is about 0.25 h^{-1}—for metal-plastic double-glazed windows, 0.4 h^{-1}—triple glazing in wooden weaves, 0.65 h^{-1}—double glazing. The analysis shows that the air exchange rate is twice as high for wooden windows. This value may actually be higher for older windows with wooden frames that have increased air permeability during operation [8].

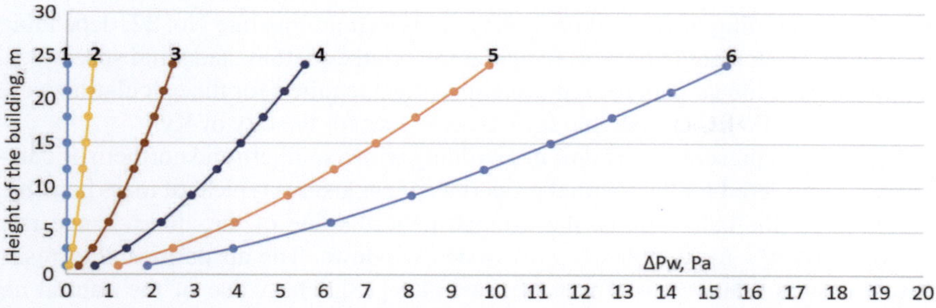

Fig. 3.8 Variation of wind pressure as a function of wind speed with a north wind direction on the north and east sides of the building (θ = 0°): 1–w = 0 m/s; 2–3 m/s; 3–6 m/s; 4–9 m/s; 5–12 m/s; 6–15 m/s

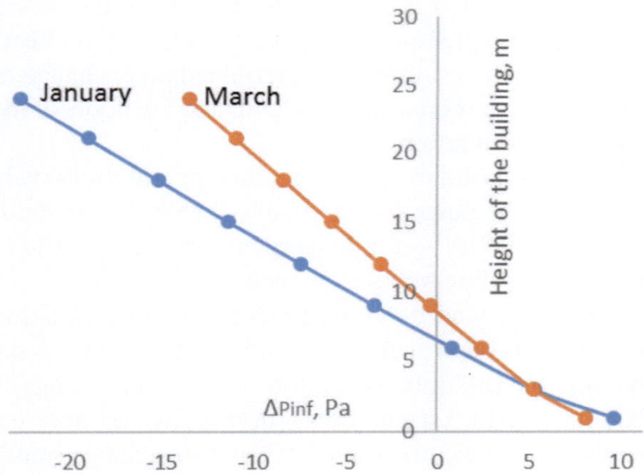

Fig. 3.9 Variation of the average monthly natural ventilation pressure by height for January and March data of a typical year for NW orientation rooms

Throughout the year, the maximum value of the air exchange rate is highest for double-glazed windows and can range from −2 to 1.2 h^{-1}, which is typical for periods with gusty winds of more than 15 m/s [8].

As mentioned above, the methodology under consideration allows taking into account both infiltration and exfiltration natural air flows in a building. For the off-season period, the air exchange rate is slightly reduced compared to the cold period, which is explained by a decrease in the temperature difference between the indoor and outdoor environment [8].

On the ground floor, the phenomenon of infiltration is observed (fresh air flows naturally through leaks in window structures), on the middle floor for all months, the multiplicity is close to zero on average, which is associated with the building's NPL (stack effect), and on the upper floors, exfiltration is observed, which corresponds to negative multiplicity values [8].

According to the empirical methodology for determining air exchange based on the improved ASHRAE methodology, the change in the pressure difference from the stack effect, wind pressure, and total pressure difference was determined for the educational building of Igor Sikorsky Kyiv Polytechnic Institute No. 22, depending on changes in such parameters as temperature, wind direction, and wind speed [15].

The hourly climate data for representative days required for the calculation were taken from the IWEC climate file for a typical year for the city of Kyiv.

A group of representative rooms in a building with a southern and northern orientation was considered for the thermal properties of enclosures typical of mass housing. According to the calculations, the average natural value of the air exchange rate is about 0.65 h^{-1} for double-glazed wooden windows. The air permeability resistance for such windows is 0.5 (m^2-h-$Pa^{2/3}$)/kg [15]. The value of the natural air exchange rate may actually be higher for older windows with wooden frames that have increased air permeability during operation.

The calculations took into account that the building maintains a constant indoor temperature of 18 °C. Since there is no forced ventilation in the research facility, the total pressure was determined as the sum of the pressure difference from the stack effect and the pressure difference from the wind effect.

The hourly results of the determined air exchange rate are shown in Fig. 3.10 for representative rooms on the 1st, 4th and 7th floors of the building of the northern and southern orientations.

For the considered type of window structures, the values of the natural air exchange rate are in the range from 0.58 to −0.85 h^{-1}—for the southern orientation and from 0.59 to −0.76 h^{-1}—for the northern orientation. The largest range of fluctuations in this value is observed for the month of January.

On the ground floor, the phenomenon of infiltration is observed (fresh air flows naturally through leaks in window structures), on the middle fourth floor for all months the multiplicity is close to zero on average, which is associated with the building's NPL (stack effect), on the seventh floor exfiltration is observed, which explains the negative values of the air exchange multiplicity. In other words, air is being sucked in on the ground floor, and expelled on the seventh floor.

The graphs show that the value of the air exchange rate is variable, due to the dynamic change in ambient temperature and wind speed. For example, on 21 January, the air exchange rate in the rooms of the 7th floor on the north side was 6.2 h^{-1}, this value is due to the average wind speed for the day of 10.3 m/s and the westerly wind direction. The relatively stable air exchange rate in November in the rooms on the 1st and 7th floors on the north side averaged 0.6 h^{-1}, which was due to slight fluctuations in temperature and wind speed during the month, as well as prevailing east and west wind directions.

The change in the air exchange rate for different air permeability resistance of translucent enclosures was studied. It follows from the calculations that windows with improved thermal and physical characteristics are less sensitive to fluctuations in environmental parameters. In general, the range of changes in the natural air exchange rate is 0–0.25 h^{-1} for double-glazed windows with PVC profiles, 0–0.5 h^{-1} for windows with triple glazing in wooden frames, and 0–0.85 h^{-1} for double glazing.

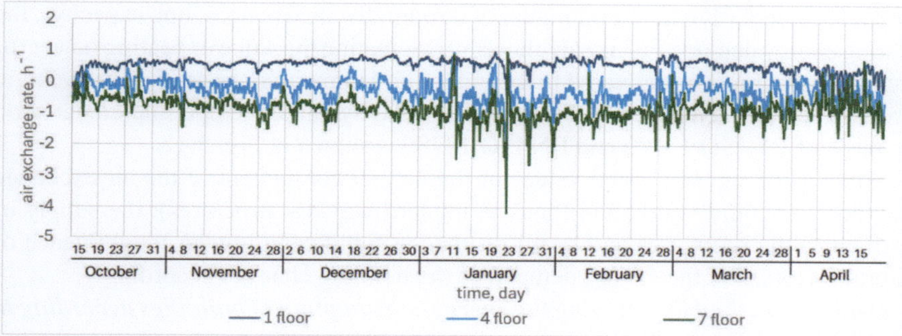

a)

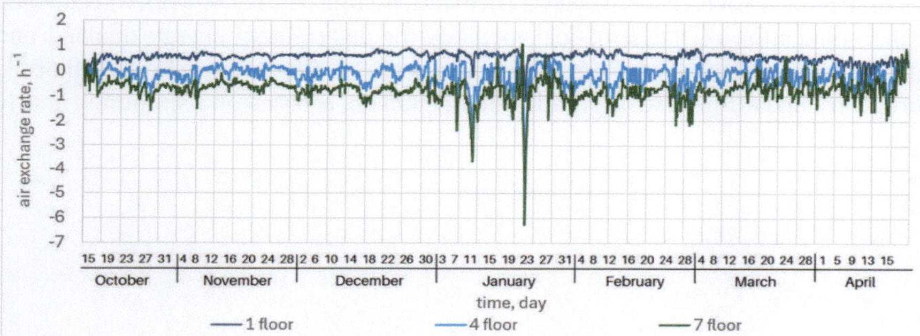

b)

Fig. 3.10 Changes in air exchange rate by month during the heating period for northern (**a**) and southern (**b**) orientations

3.4 Calculation of Energy Requirements and Analysis of Load Fluctuations When Changing the Natural Level of Air Exchange Rate

The group of representative rooms are mass-produced buildings from the 1970s. The dimensions of the room are 5.5 × 6.1 m, the height of the room is 3 m. The room has one outer wall (5.5 × 3.0 m) with a window (5 × 2 m). The outer wall has a thermal resistance of 0.8 m^{2o} S/W (the main layer is single-brick masonry). The ceilings above the heated rooms are reinforced concrete—20 cm. The building is located in Kyiv. The design internal air temperature is 18 °C. The solar heat gain coefficient of translucent structures in the room area is 0.56. Climatic hourly data for the city of Kyiv were used from the IWEC International Weather File [11]. Solar heat gain into the room zone was calculated using the IWEC file data based on the Full interior and exterior with reflection methodology, which takes into account reflection from surfaces and soil and other equally important factors.

Figure 3.11 shows the hourly variation of the outdoor temperature, air exchange rate and solar heat gain for the NW and SE orientations of the room. According to the 5R1C model (dynamic grid model of a room, five resistances, one capacity), the heating energy demand F_{HC} is calculated by calculating the value of heating power for each hour that must be delivered to the indoor air temperature node (θ_{air}) to maintain a certain setpoint temperature. The scheme of the dynamic method implemented on the basis of EN ISO 13790 and EN ISO 13786 is shown in Fig. 3.12.

When using the calculated values of the natural component of the air exchange rate, due to natural ventilation (gaps in modern window structures), the supply of fresh air to the room area and heat loss change significantly under the influence of natural factors, and the energy demand of the building changes accordingly.

Dynamic method for studying the energy consumption of buildings according to EN ISO 13790. For a detailed analysis of energy characteristics, it is advisable to use dynamic models for calculating energy demand. A large number of works are devoted to the application of mathematical methods for determining energy demand and establishing the thermal efficiency of buildings [7, 16–21]. Another approach of the European standard [22], which is adopted in Ukraine, is based on a simplified hourly

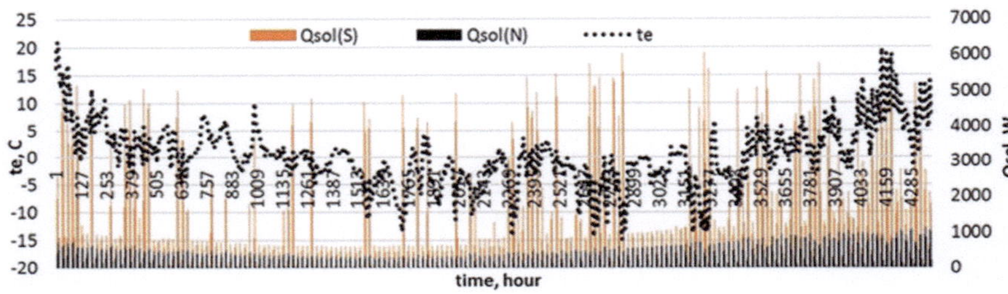

Fig. 3.11 Changes in outdoor air temperature t_{out}, solar heat gain to the room zone Q_{sol} for the northern (N) and southern (S) orientation of the room

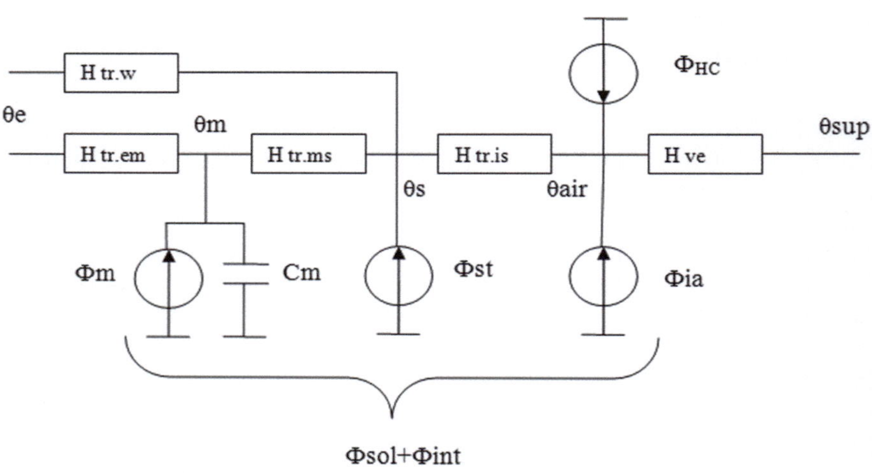

Fig. 3.12 Five resistances, one capacitance (5R1C) model

method of calculating energy demand. The standard [22] proposes a grid model—five resistances, one capacitance (5R1C), which allows the model to be implemented in a simplified three-node hourly method. This approach requires writing or using existing programs to implement this method. Figure 3.12 shows a simplified diagram of the method implementation.

The heating energy demand is calculated by calculating the value of the heating or cooling capacity for each hour, $\Phi_{HC.nd}$ that must be supplied to, or taken from, the indoor air temperature node θ_{air}, to maintain a certain minimum or maximum setpoint temperature. Operating temperature θ_{op} is the weighted average of the air temperature and the average radiation temperature.

Heat transfer by ventilation, H_{ve} is directly related to the indoor air temperature node θ_{air}, and the node representing the supply air temperature, θ_{sup}. The heat transfer by the transmission is divided into a part through the translucent openings, $H_{tr.w}$ assumed to have zero thermal mass, and the remainder through H_{op} with thermal mass, which in turn is divided into two parts: $H_{tr.em}$ and $H_{tr.ms}$. The solar and internal heat inputs are distributed between the internal air temperature node, θ_{air}, the central node, θ_s (a mixture of θ_{air} and average radiation temperature θ_r) and a node representing the building zone array, θ_m. The thermal mass is represented by the specific heat capacity, C_m defined between $H_{tr.ms}$ and $H_{tr.em}$. The conduction coupling is defined between the indoor air temperature node and the central node. The heat flux due to internal sources, Φ_{int} and the heat flux due to the sun, Φ_{sol} are divided between three nodes: the indoor air temperature node θ_{air}, and the internal nodes θ_s, θ_m. This scheme can be implemented on the basis of the EN 13,790 and EN 13,786 standards [22, 23]. The EN 13,786 standard [23] allows the calculation of the inertial characteristics of enclosures, in contrast to the generalized approaches proposed in the standard [22]:

$$H_{tr.is} = h_{is}A_{tot}; \tag{3.17}$$

$$H_{tr.ms} = h_{ms}A_m; \tag{3.18}$$

$$H_{tr.em} = \frac{1}{\frac{1}{H_{op}} - \frac{1}{H_{tr.ms}}}; \tag{3.19}$$

$$H_{tr.1} = \frac{1}{\frac{1}{H_{ve}} + \frac{1}{H_{tr.is}}}; \tag{3.20}$$

$$H_{tr.2} = H_{tr.1} + H_{tr.w}; \tag{3.21}$$

$$H_{tr.3} = \frac{1}{\frac{1}{H_{tr.2}} + \frac{1}{H_{tr.ms}}}; \tag{3.22}$$

$$C_m = \sum k_j A_j; \tag{3.23}$$

$$\Phi_{ia} = 0.5\Phi_{int}; \tag{3.24}$$

$$\Phi_m = \frac{A_m}{A_{tot}}(0.5\Phi_{int} + \Phi_{sol}); \tag{3.25}$$

$$\Phi_{st} = \left(1 - \frac{A_m}{A_{tot}} - \frac{H_{tr.w}}{9,1A_{tot}}\right)(0.5\Phi_{int} + \Phi_{sol}); \tag{3.26}$$

$$\Phi_{m.tot} = \Phi_m + H_{tr.em}\theta_e + \frac{H_{tr.3}(\Phi_{st} + H_{tr.w}\theta_e + H_{tr.1}(\frac{\Phi_{ia}+\Phi_{HC.nd}}{H_{ve}} + \theta_{sup})}{H_{tr.2}}; \tag{3.27}$$

$$\theta_m = \frac{\theta_{m,t} + \theta_{m,t-1}}{2}; \tag{3.28}$$

$$\theta_{m,t} = \frac{\theta_{m,t-1}(\frac{C_m}{3600} - 0.5(H_{tr.3} + H_{tr.em}) + \Phi_{m.tot}}{\frac{C_m}{3600} + 0.5(H_{tr.3} + H_{tr.em})}; \tag{3.29}$$

$$\theta_s = \frac{H_{tr.ms}\theta_m + \Phi_{st} + H_{tr.w}\theta_e + H_{tr.1}(\frac{\Phi_{ia}+\Phi_{HC.nd}}{H_{ve}} + \theta_{sup})}{H_{tr.ms} + H_{tr.w} + H_{tr.1}}; \tag{3.30}$$

$$\theta_{air} = \frac{H_{tr.is}\theta_s + H_{ve}\theta_{sup} + \Phi_{ia} + \Phi_{HC.nd}}{H_{tr.is} + H_{ve}}, \tag{3.31}$$

where:

θ_{air}—internal air temperature, °C;

θ_s—temperature of the central node s, °C;

θ_m—temperature at node m, °C;

θ_{sup}—supply air temperature of ventilation systems, °C;

Φ_{int}—heat input from internal sources, W;

Φ_{sol}—heat flux from the sun, W;

Φ_{ia}, Φ_m, Φ_{st}—heat flux from internal sources and the sun is distributed among 3 nodes, θ_{air}, θ_s, θ_m;

$\Phi_{m.tot}$—total heat flux, W;

$\Phi_{HC.nd}$—heat flow from heating or cooling, W;

h_{is}—heat transfer coefficient between the indoor air temperature node, θ_{air} and the central node, θ_s, with a fixed value$h_{is} = 3.45\frac{W}{m^2K}$;

h_{ms}—heat transfer coefficient between nodes m and s, with a fixed value $h_{ms} = 9.1\frac{W}{m^2K}$;

A_m—effective mass area, m^2;

A_j—area of the jth element, m^2;

A_{tot}—area of all external fences of the building zone, m^2;

C_m—internal heat capacity, J/K;

k_j—internal heat capacity per unit area of the jth building element, J/(m^2·K);

$H_{tr.ms}$—conductivity coupling between nodes m and s, W/K;

$H_{tr.em}$—conductivity coupling between nodes m and the external environment, W/K;

$H_{tr.is}$—conductivity coupling between the nodes s and the internal air, W/K;

H_{op}—generalised heat transfer coefficient of opaque building elements, W/K;

$H_{tr.w}$—heat transfer coefficient of transparent building elements, W/K;

$H_{tr.1}$, $H_{tr.2}$, $H_{tr.3}H_{tr.3}$—conductivity of conventional nodes 1, 2, 3 W/m;

H_{ve}—total heat transfer coefficient by ventilation, W/K.

The model is implemented in the Mathcad environment.

The mathematical modelling of the specific energy demand of representative rooms of the northern (N) and southern (S) orientation was carried out, taking into account only transmission losses ($n = 0$), for the normative values of the air exchange rate ($n = 1\ \text{h}^{-1}$) and for the calculated value of the natural component ($n = var$) of air exchange.

Figure 3.13 shows the annual specific energy demand of a building for heating for the north (N) and south (S) orientation, using representative ground floor rooms with double-glazed energy-saving PVC double-glazed windows. The maximum estimate of additional costs for energy consumption due to natural ventilation (infiltration) compared to the energy consumption of the building with transmission losses only ($n = 0$) differs by 23% for northern and 43% for southern orientations.

The calculated value of the natural component of air exchange does not allow to ensure the standard level of air exchange for comfortable working conditions. A comparison of the difference in specific energy demand calculated for the normative values ($n = 1\ \text{h}^{-1}$) with the calculated level of natural air exchange was carried out and it was found that the difference in energy demand for the heating period can be more than 50% compared to the baseline. For example, on the ground floor of NW-orientated premises, the difference is 50%, and on the SE—65%. This is the

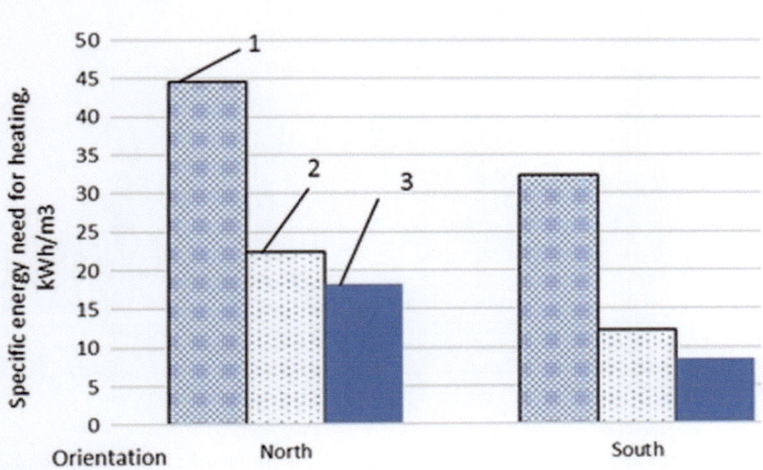

Fig. 3.13 Annual specific energy demand for heating: 1—at standard values of air exchange rate $n = 1\ \text{h}^{-1}$; 2—at calculated (natural component) air exchange; 3—in the absence of air exchange

maximum possible value of savings under actual air exchange conditions compared to the normative ones. These savings can be reduced by increasing air exchange during occupancy hours, for example, through ventilation.

Figure 3.14 shows the change in the specific load on the heating system for the normative (1) and calculated (2) values of the natural air exchange rate and for a room with only the transmission component of losses (3), oriented to the north (a) and south (b), located on the ground floor.

According to the results of mathematical modelling, it was found that at a certain level of natural air exchange for the type of window under consideration (air permeability resistance), the heating system of south-facing rooms can be switched off during the off-season (October, April), i.e. the window works as a passive heating system.

During the cold season, the energy demand calculated for the normative air exchange rate and the one based on variable environmental conditions (natural air exchange rate) differs by up to 50%, and for the off-season by up to 75%.

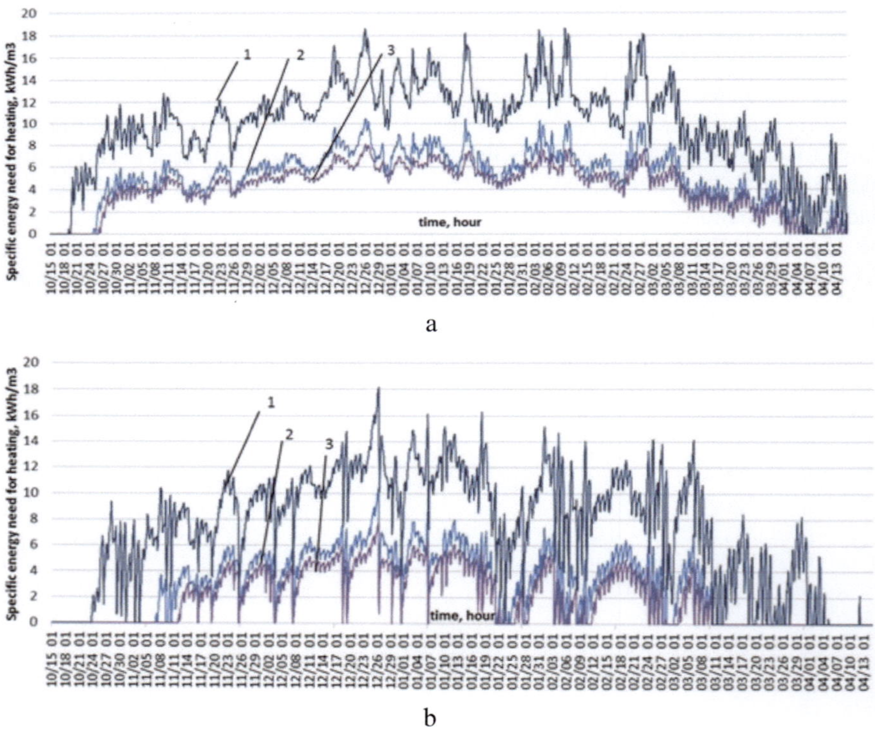

Fig. 3.14 Heating system load for the heating period of a north-facing room (**a**) and south-facing room (**b**): 1—at standard values of air exchange rate $n = 1\ \text{h}^{-1}$; 2—at calculated values (natural component); 3—in the absence of air exchange

For south-facing rooms, the fluctuations in the heating load during the day are somewhat greater than for north-facing rooms, which is explained by solar heat gain in the room area. On some days, there are also short-term heating outages during peak solar activity hours.

3.5 Conclusions

An improved empirical methodology for determining the natural air exchange rate based on the improved ASHRAE methodology is proposed. An hourly calculation of the natural component of the air exchange rate for the heating period of a representative group of premises was carried out and its impact on energy consumption was investigated.

As a result of the study, the influence of the components of the air exchange rate, which depend on the difference in wind and stack effect pressures, was determined. For the climatic conditions of the city of Kyiv, it was found that for single-chamber double-glazed windows in paired wooden weaves, the air exchange rate for a representative group of rooms is –0.3...0.3 h^{-1}, which is significantly less than the minimum standard value for educational institutions, and the results are in line with the research of foreign scientists.

The chapter proposes an improved hourly dynamic grid model for determining the energy demand of representative premises, which is based on the EN series standards, namely DSTU B EN ISO 13790:2013 and EN ISO 13786:2007.

It has been established that when calculating the energy demand hourly using the standard value of $n = const = 1\ h^{-1}$ and $n = var$, the calculated value leads to the shutdown of the heating system during the off-season (October, April), as well as to a difference between the baseline and the actual energy consumption of more than 50% during the heating period.

References

1. Jalaei, F., Jrade, A.: Integrating building information modeling (BIM) and energy analysis tools with green building certification system to conceptually design sustainable buildings. J. Inf. Technol. Constr. **19**, 494–519 (2014)
2. Harish, V.S.K.V., Kumar, A.: A review on modeling and simulation of building energy systems. Renew. Sustain. Energy Rev. **56**, 1272–1292 (2016)
3. Bilous, I.Y., Deshko, V.I., Sukhodub, I.O.: Building energy modeling using hourly infiltration rate. Mag. Civ. Eng. **96**(4), 27–41 (2020)
4. Prek, M., Butala, V.: Principles of exergy analysis of human heat and mass exchange with the indoor environment. Int. J. Heat Mass Transf. **53**, 25–26:5806–14 (2010)
5. German Association of Engineers: Calculation of transient thermal response ofrooms and buildings – modelling of rooms. 91.140.10 (VDI 6007-1), BeuthVerlagGmbH, Düsseldorf (2012)
6. Ghita, S.A., Catalina, T.: Energy efficiency versus indoor environmental quality in different Romanian countryside schools. Energy Build. **92**, 140–154 (2015). https://doi.org/10.1016/j.enbuild.2015.01.049
7. MacKay, D., Ryder-Cook, D.: Thermal Modelling of Buildings, p. 23. UIT Cambridge, Cambridge (2009). http://www.withouthotair.com
8. Deshko, V., Bilous, I., Buyak, N., Naumchuk, O.: Prospects for the use of renewable energy sources while increasing the energy efficiency level of office buildings to the level of nZEB. Rocznik Ochrona Srodowiska **25**, 148–158 (2023). https://doi.org/10.54740/ros.2023.015
9. Deshko, V., Bilous, I., Vynogradov-Saltykov, V., Shovkaliuk, M., Hetmanchuk, H.: Integrated approaches to determination of co2 concentration and air rate exchange in educational institution. Rocznik Ochrona Srodowiska **22**(1), 82–104 (2020). https://ros.edu.pl/images/roczniki/2020/007_ROS_V22_R2020.pdf
10. Fareniuk H.H., Filonenko O.I., Tymofieiev M.V.: Enerhoefektyvnist hromadskykh budynkiv z vrakhuvanniam erhonomiky teplovoho seredovyshcha, Komunalne hospodarstvo mist. Seriia: Tekhnichni nauky ta arkhitektura **135**, 119–124 (2017)
11. International Weather for Energy Calculations. https://energyplus.net/weather-location/europe_wmo_region_6/UKR
12. Buyak, N., Deshko, V., Bilous, I.: Changing energy and exergy comfort level after school thermomodernization. Rocznik Ochrona Srodowiska **23**, 458–469 (2021). https://doi.org/10.54740/ros.2021.031
13. DSTU-N B V.1.1-27: Budivelna kilmatologiya. K.: Minregion Ukraïni, 2011, 127s (2010)
14. Enerhetychna efektyvnist budivel. Metod rozrakhunku enerhospozhy-vannia pid chas opalennia, okholodzhennia, ventyliatsii, osvitlennia ta hariachoho vodo-postachannia, DSTU 9190:2022, DP NDIBK, Kyiv (2022)
15. Bilous, I.Y.: Otsiniuvannia enerhetychnoi efektyvnosti budivel v umovakh dynamichnoi zminy kharakterystyk seredovyshcha. Dysertatsiia doktora tekhnichnykh nauk, Igor Sikorsky Kyiv Polytechnic Institute, Kyiv (2019)
16. Lie, B.: Improved model for solar heating of buildings. In: SIMS, 07–09 October 2015, Linköping, Sweden, pp. 299–308 (2015)
17. Harish, V.S.K.V., Kumar A.: A review on modeling and simulation of building energy systems. Renew. Sustain. Energy Rev. **56**, 1272–1292 (2016)
18. Pereira W., Bögl A., Natschläger T.: Sensitivity analysis and validation of an EnergyPlus model of a house in Upper Austria. Energy Procedia **62**, 472–481 (2014)
19. Stefanovic A., Gordic D.: Modeling methodology of the heating energy consumption and the potential reductions due to thermal improvements of staggered blockbuildings. Energy Build. **125**, 244–253 (2016)
20. Lu S., Zheng S., Kong X.: The performance and analysis of office building energy consumptionin the west of Inner Mongolia Autonomous Region, China. Energy Build. **127**, 499–511 (2016)

21. Braulio-Gonzalo M., Juan P., Bovea M.D., Rua M.J.: Modelling energy efficiency performance of residential building stocks based on Bayesian statistical inference. Environ. Modell. Softw. **83**, 198–211 (2016)
22. EN ISO 13790: Energy performance of buildings – Calculation of energy use for space heating and cooling. CEN. European Committee for Standardization, 53r (2008)
23. EN ISO 13786: Thermal performance of building component - Dynamic thermal characteristics - calculation methods. CEN. European Committee for Standardization, 27p. (2007)

Chapter 4
Experimental Studies of Air Exchange Parameters and Microclimatic Characteristics

4.1 Parametric Analysis of the Influential Components on the Level of Air Exchange

The study analysed the change in CO_2 concentration in the room depending on the change in CO_2 concentration in the supply air, the volume of supply air and human emissions.

A living room was chosen as a representative room for the analysis of the influencing factors. Figure 4.1 shows the results of calculations of carbon dioxide concentration depending on the concentration of CO_2 in the supply air, in which the parameter—the initial concentration in the room—varied from 420 to 2500 ppm (4.50 g/m^3). To demonstrate the calculations, three cases were chosen when the initial concentration of CO_2 in the supply air was at the level of the background—420 ppm (0.76 g/m^3), 1300 ppm (2.34 g/m^3), 2500 ppm (4.50 g/m^3). The volume of supply air in Fig. 4.1 (a) is 10 m^3 /h, the air exchange rate in the room in this case is 0.23 h^{-1}, in Fig. 4.1 (b)—the volume of supply air is 30 m^3 h (air exchange rate is 0.69 h^{-1}), in Fig. 4.1 (c)—the volume of supply air is 50 m^3/h (air exchange rate is 1.15 h^{-1}). The component of income from people is absent [1].

From the analysis of the data obtained from the calculations illustrated in Fig. 4.1, it follows that the rate of change in the carbon dioxide concentration in the room increases in proportion to the increase in the volume of supply air. This means that the more air that enters the room, the faster the CO_2 concentration changes. The range of change in CO_2 concentration is also affected by the difference between the initial CO_2 concentration in the room and the CO_2 concentration in the supply air [1].

Figure 4.2 shows the change in the concentration of CO_2 in a living room depending on the volume of supply air entering the room under conditions of dynamic changes in the concentration of CO_2 in the supply air. For case I, the concentration of CO_2 in the supply air varied from 2500 ppm (4.50 g/m^3) to 1700 ppm (3.06 g/m^3), for case II, on the contrary, it increased from 1700 to 2500 ppm [1].

A. Zaporozhets et al., *Buildings Energy Audit*, Studies in Systems, Decision and Control 641, https://doi.org/10.1007/978-3-032-07200-9_4

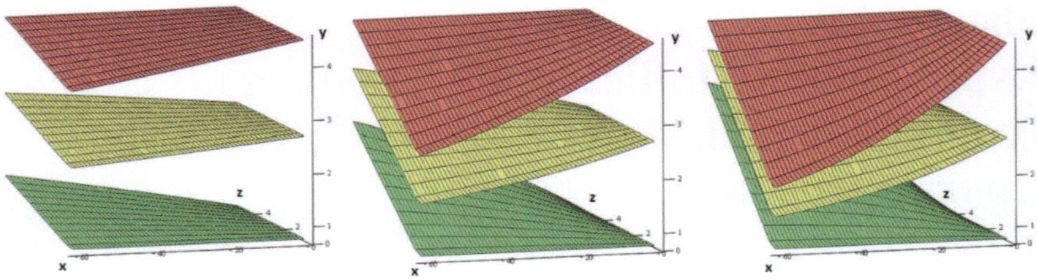

(a) (b) (c)

Fig. 4.1 Change in carbon dioxide concentration depending on the concentration of CO_2 in the supply air for conditions **a** supply air volume 10 m^3/h; **b** supply air volume 30 m^3/h; **c** supply air volume 50 m^3/h: x-axis—time, minutes; y-axis—concentration of CO_2 in the room air, g/m^3; z-axis is the concentration of CO_2 in the supply air, g/m^3; —the initial concentration of CO_2 in the room air is 420 ppm —the initial concentration of CO_2 in the room air is 1300 ppm; —initial concentration of CO_2 in the room air is 2500 ppm

For the conditions where the initial CO_2 concentration in the supply air is 4.50 g/m^3 and is significantly higher than the initial CO_2 concentration in the room of 1.08 g/m^3 (Fig. 4.2a (Case II)), the CO_2 level in the room initially increased, but after a while reached its peak and also began to decrease. For the conditions shown in Fig. 4.2a (case I), the CO_2 level in the room will decrease because the room will receive air with a lower concentration. The opposite cases, when the concentration of CO_2 in the supply air is dynamically increasing, and the level of CO_2 in the room increases accordingly, are shown in Fig. 4.2b.

The most influential factor on air quality is the presence of occupants and their characteristics. People are the largest source of CO_2 in a room. To analyse the relationship between CO_2 concentration and air exchange rate, it is necessary to calculate the rate of CO_2 formation from human breathing. ASHRAE Handbook of Fundamentals [2] and ASTM D6245-18 [3] describe the estimation of CO_2 emissions as follows. According to the ASHRAE Handbook, the empirical formula for calculating the CO_2 emission factor (l/s) can be expressed as follows:

$$\mathrm{V}_{\mathrm{CO_2}} = \mathrm{RQ} \cdot \frac{0.0055887\mathrm{W}^{0.425} H^{0.725} M}{0.23\mathrm{RQ} + 0.77}, \tag{4.1}$$

where RQ is the respiratory quotient, the volume ratio of formed CO_2 to consumed oxygen (O_2), dimensionless (respiratory quotient is 0.83 for light or sedentary activities ($M < 1.5$ met, 1 met = 58.1 W/m^2)); W—body weight (kg); H—body height (m); M—metabolic rate (met), depending on the level of physical activity.

For example, for an adult of average height ($H = 1.8$ m), weighing $W = 85$ kg, engaged in sedentary work at $M = 1.2$ met, the corresponding CO_2 production rate is 0.006 l/s, which is 42.57 g/h. These are the parameters of an average person used in the calculations for this study. In [4], based on annual observations, the behaviour of residents as a key factor affecting the air exchange rate was investigated.

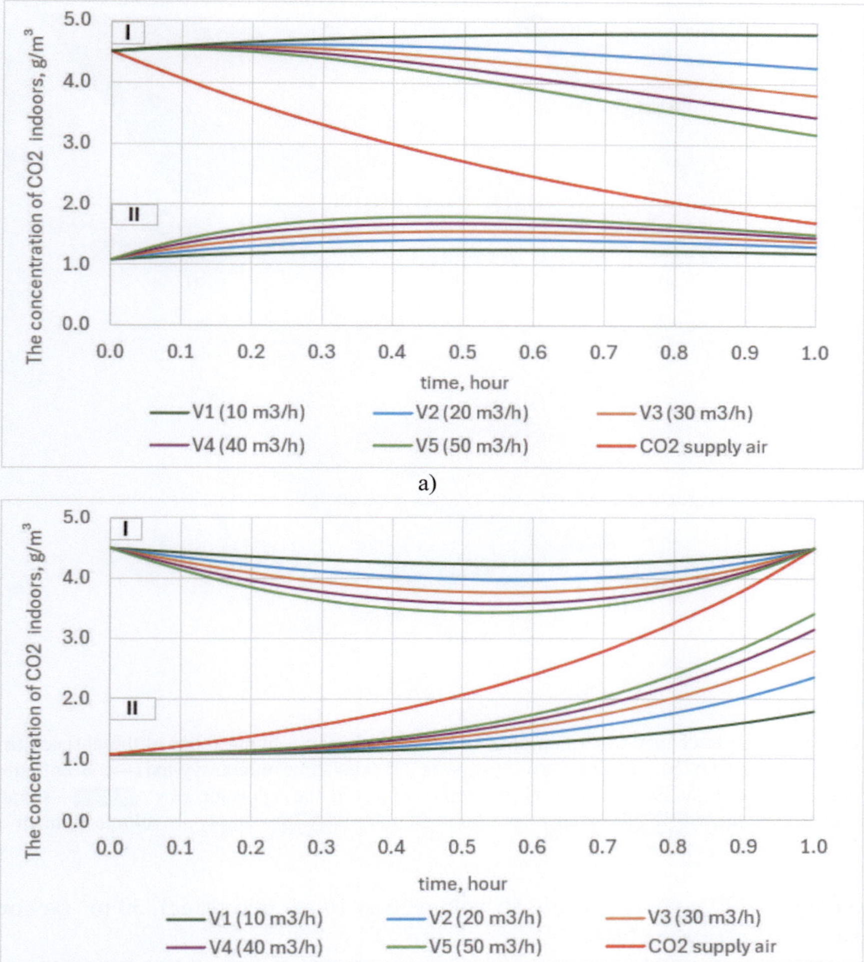

a)

b)

Fig. 4.2 Changes in carbon dioxide concentration over time under conditions of dynamic changes in the concentration of CO_2 in the supply air with a decrease in the concentration of CO_2 in the supply air (**a**) and an increase in the concentration of CO_2 in the supply air (**b**)

Thus, the presence of people in the premises has the greatest impact on the change in CO_2 concentration in the room, as shown in Fig. 4.3.

Accordingly, with the increase in the number of people present in the room, a larger volume of supply air is required to ensure appropriate microclimate conditions. For this series of calculations, the range of 2500–420 ppm of the starting CO_2 concentration was selected, the value in the external supply air was assumed to be

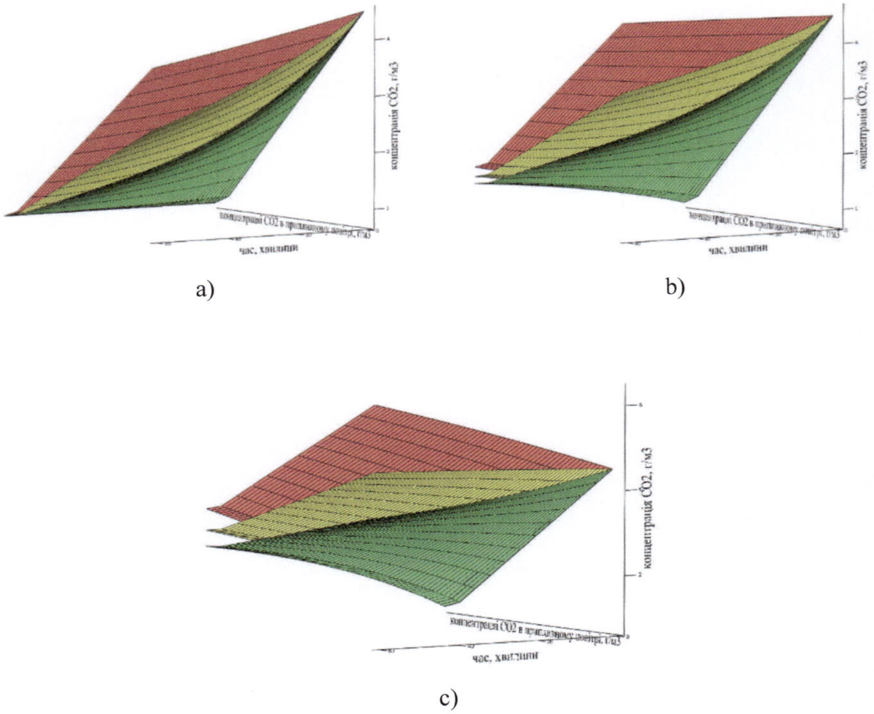

Fig. 4.3 Dynamic change of carbon dioxide concentration in a living room in the absence (**a**), presence of 1 person (**b**) and presence of 3 people (**c**): x-axis—time, minutes; y-axis—concentration of CO_2 in the room air, g/m^3; z-axis—concentration of CO_2 in the supply air, g/m^3; [green]—supply air volume 50 m^3/h; [yellow]—supply air volume 30 m^3/h; [red]—supply air volume 10 m^3/h

constant at 420 ppm, the supply air volume was 10 m^3 (red plane), 30 m^3 (yellow plane), 50 m^3 (green plane).

According to DBN V.2.5-67:2013 [5] and DSTU B EN 15,251:2011 [6], the recommended level of carbon dioxide in the premises of existing buildings should not exceed 800 ppm above the outdoor level, i.e. a typical value of 1200 ppm.

For a room of 17.4 m^2 in which 3 adults are engaged in sedentary work, it is necessary to ensure an air exchange rate of 2.01 h^{-1} so that the level of CO_2 in the air does not exceed the maximum permissible level.

4.2 Integrated Approach to Determining CO_2 Concentration and Air Exchange in Educational Institutions

In this chapter, full-scale and simulation modelling was carried out to determine the change in CO_2 concentration in the premises and natural air exchange in educational institutions, using the example of facilities located in the city of Kyiv, the research scheme is shown in Fig. 4.4.

The research was conducted in stages and developed in three schools and a university. All schools are typical, built in the years of mass construction, H-type. School 1 was used to justify the selection of representative rooms, investigate the local location of sensors in the premises to determine the CO concentration$_2$ and other input parameters for calculating the natural air exchange rate. In school 2, CO_2 concentrations were measured in representative rooms at the beginning and end of classes. In school 3, CO_2 concentrations were measured for two variants: (1) measurements of CO_2 concentration in representative rooms at the beginning and end of classes; (2) recording the dynamics of changes in CO_2 concentration in representative rooms in 10-min increments, taking into account changes in the number of students, physical

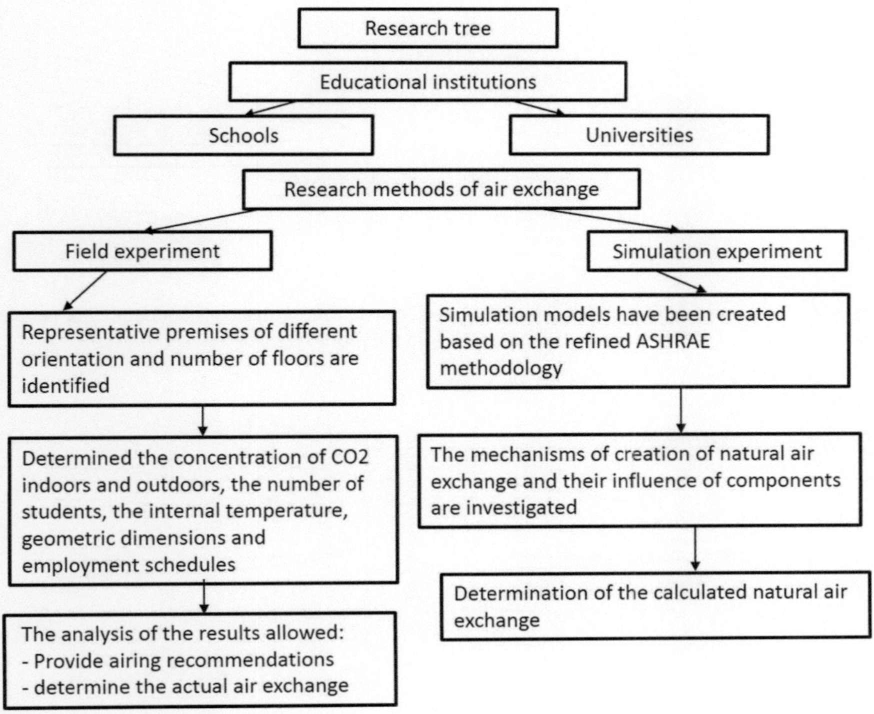

Fig. 4.4 Scheme of the research

activity and classroom occupancy schedule. Experimental determinations of changes in concentration in representative rooms were carried out in the educational complex of Igor Sikorsky Kyiv Polytechnic Institute. A local study of the background external concentration of CO_2 was carried out near the objects under consideration during the day with different distances from them.

A computational study of the impact of air exchange, initial concentration and emissions of CO_2 in the room, air flow by infiltration and exfiltration on the concentration of CO_2 in the room was carried out (Fig. 4.5).

The decreasing (Fig. 4.5b)/increasing (Fig. 4.5a) exponential curve at the same m/V ratios is due to different initial CO_2 concentrations and indicates the time of ventilation/no ventilation, respectively.

Figure 4.6 shows the effect of the number of people present on the CO2 concentration in the room at constant air exchange: 0.5 h^{-1} (Fig. 4.6a) and 1 h^{-1} (Fig. 4.6b) at a starting concentration of $C_{i0} = 2$ g/m^3. A decrease in the air exchange rate leads to a smaller bend in the exponential dependence of the CO_2 concentration in time. The number of people present has a more significant effect on the CO_2 concentration in the room (Fig. 4.5) than the air exchange rate (Fig. 4.6).

Experimental part. The experimental determination of the background concentration of CO_2 during the day locally near the considered objects showed that the

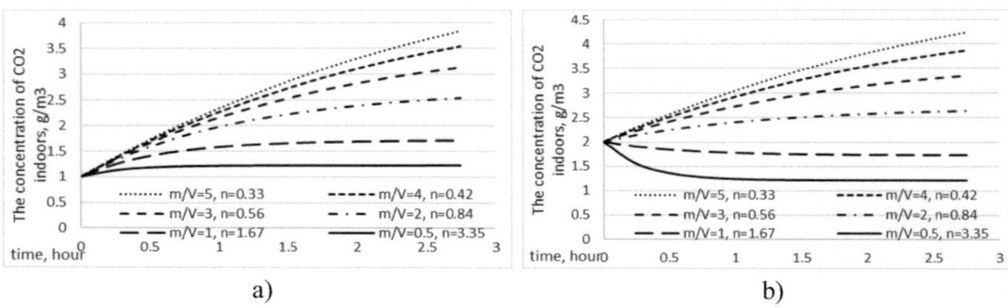

Fig. 4.5 Changes in the internal concentration of CO_2 over time with a constant number of people present and different initial concentrations of CO_2 in the room

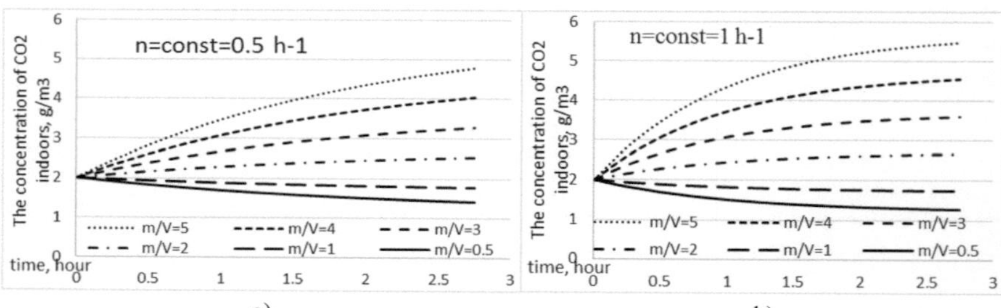

Fig. 4.6 Changes in the internal CO_2 concentration over time with constant air exchange and different numbers of people present (CO_2 emissions in the room)

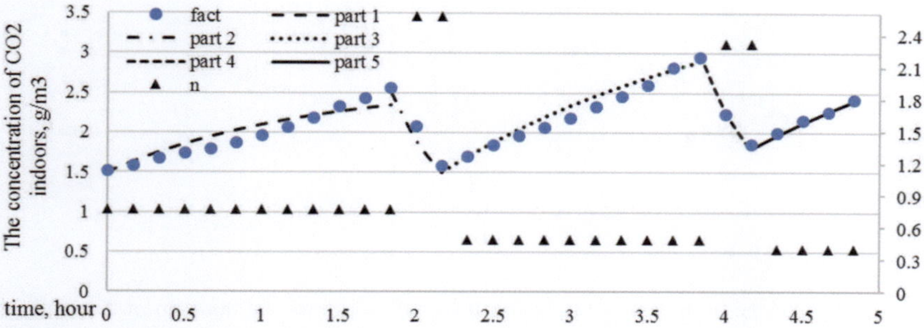

Fig. 4.7 Changes in the internal CO_2 concentration and air exchange rate in the classroom of school 3

background concentration of CO_2 is in the range of 400–420 ppm. On the example of school 1, an experimental study was conducted to determine local measurements of CO_2 concentration in the premises during and after classes. Measurements were made of the distribution of CO_2 concentration after classes over the classroom area, according to which the difference between the values at the level of desks was 30…180 ppm. The studies showed that it is acceptable to use the integral characteristic of CO concentration$_2$ in a representative room in the centre of the room at a height of about one metre (working area). Measurements of indoor CO_2 concentration in school 2 during the holiday period were quite close to the background concentration, with the exception of classes where clubs were held. During the study period, control measurements of CO_2 concentrations were made indoors at the beginning of classes, after classes, and after a break [1].

It was found that the CO_2 concentration of varies during the lesson by 700–1100 ppm and significantly exceeds the norm. During the break, the concentration of CO_2 decreases by 500–1000 ppm, depending on the type of ventilation [7].

In School 3, 10 min of CO_2 concentration measurements were made in representative classrooms. Figure 4.7 shows the determination of the air exchange rate for the example of high school during three classes and two breaks [7].

The air exchange rate during the class period (part 1, 3, 5) is in the range of 0.5…0.75 h^{-1} (closed windows and doors). During breaks, students leave the room and the classrooms are ventilated by opening windows and doors, during this period (part 2.4) the air exchange rate increases to 2.9–3.5 h^{-1}. For the range under consideration, the weighted average air exchange rate is 0.93 h^{-1} and even with forced ventilation of classrooms is insufficient in terms of ensuring the proper level of air exchange and permissible CO_2 concentration [7].

Similar studies of changes in CO concentration$_2$ were conducted for the academic building of Igor Sikorsky Kyiv Polytechnic Institute. The building is a long 7-storey building with a technical floor, the main part of classrooms and laboratories has one external enclosure oriented to the S or N. In the building, the classrooms are ventilated by opening the doors to the corridors, mechanical ventilation does not work. Using

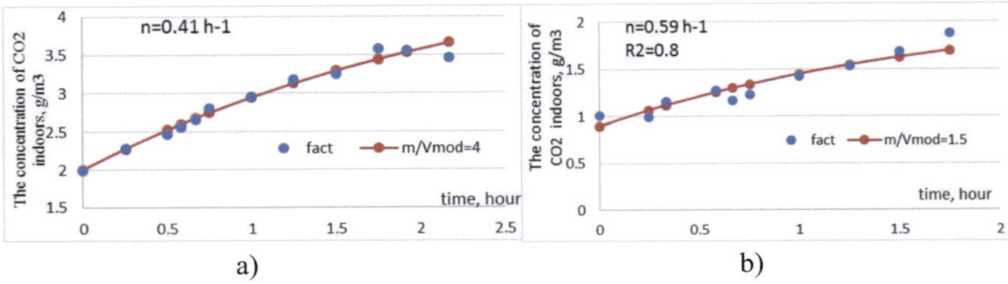

Fig. 4.8 Air exchange in the rooms on the 3rd floor (**a**) and on the 1st floor (**b**) of the southern orientation

the example of processing experimental data on changes in CO_2 concentration in classrooms, the natural air exchange rate was calculated (Fig. 4.8) [7].

The decrease in the air exchange rate on the 3rd floor is due to a decrease in the stack effect (in accordance with the components of the air exchange mechanism).

Similar studies were conducted for large lecture halls, where the impact of ventilation during breaks was more pronounced. Figure 4.9 shows the determination of the air exchange rate based on the average air exchange (a) and broken down into periods when the doors were open (parts 2, 4) and closed (parts 1, 3) (b) [7].

The analysis shows that the value of the average air exchange is 0.48 h^{-1} (processing of the graph with average V values), with separate processing of parts 2, 4 and part 1, 3, it was found that the air exchange rate during periods of closed doors is 0.4–0.6 h^{-1}, with the doors open it can reach 0.95 h^{-1}. The weighted average air exchange rate for the considered period of time in the piecewise analysis is 0.43 h^{-1}, in contrast to 0.48 h^{-1} for the averaged equation [7].

Simulation modelling. For the Igor Sikorsky Kyiv Polytechnic Institute academic building, simulation modelling of the natural air exchange of north and south orientation rooms was carried out.

Figure 4.10 shows the change in climatic conditions over the three days during which the experimental studies were conducted.

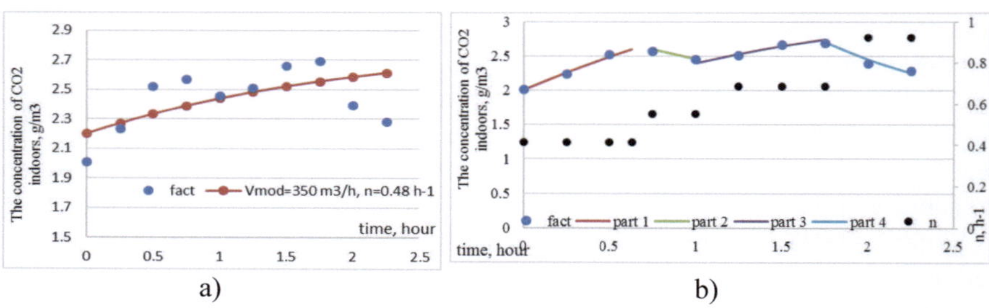

Fig. 4.9 Changes in CO_2 concentration in the lecture hall of Igor Sikorsky Kyiv Polytechnic Institute

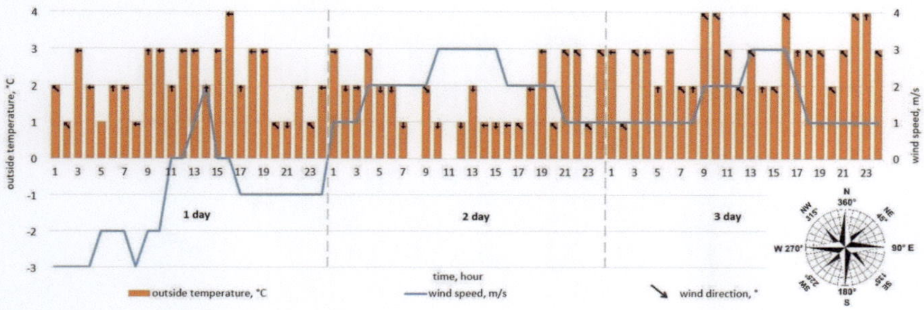

Fig. 4.10 Changing climate conditions

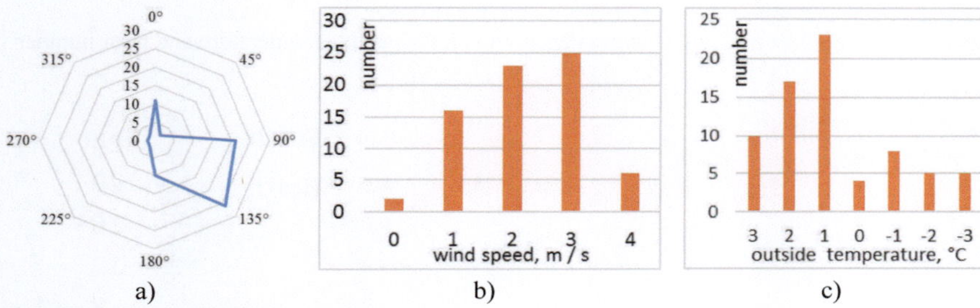

Fig. 4.11 Preferred external weather conditions: **a**—wind direction, **b**—wind speed, **c**—temperature

The analysis of meteorological data (Fig. 4.11) shows that the prevailing wind direction is Southeast (135°) (Fig. 4.11a), wind speed is 3 m/s (Fig. 4.11b), and the outside temperature is minus 1 (Fig. 4.11c).

The pressure change under the influence of the stack and wind effect was studied for the prevailing weather conditions. Figure 4.12 shows the change in the pressure difference of the stack effect depending on the floor and external temperature at a constant internal air temperature of 18 °C.

The pressure change due to the stack effect is directly proportional to the temperature difference between the outside and inside of the building. Negative values of ΔPs are due to the reference point of the pressure difference from the neutral pressure (NPL), which is located on the middle floor of the building. The NPL level is the level at which the internal and external pressures are equal. From the NPL level towards the ground floor, the pressure difference is positive, and towards the top floor it is negative.

Figure 4.13 shows the wind effect pressure profiles for the north and south orientation of the building for the prevailing wind direction and speed. For the southern orientation of the premises, the wind pressure profile has a positive sign, which is due to the positive value of the coefficient C_p (θ), which depends on the wind speed.

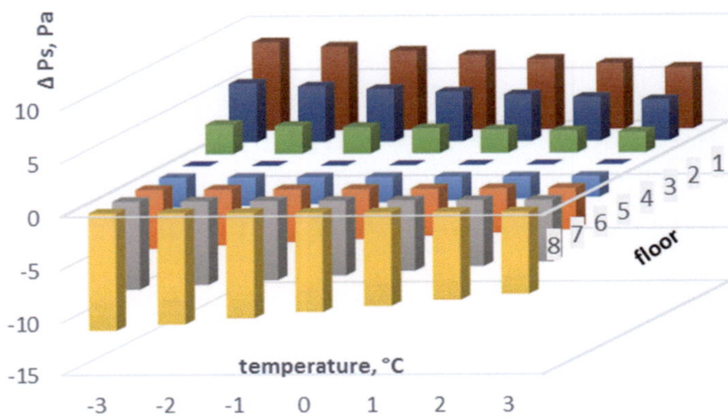

Fig. 4.12 Stack effect pressure changes depending on the outside temperature and floor number

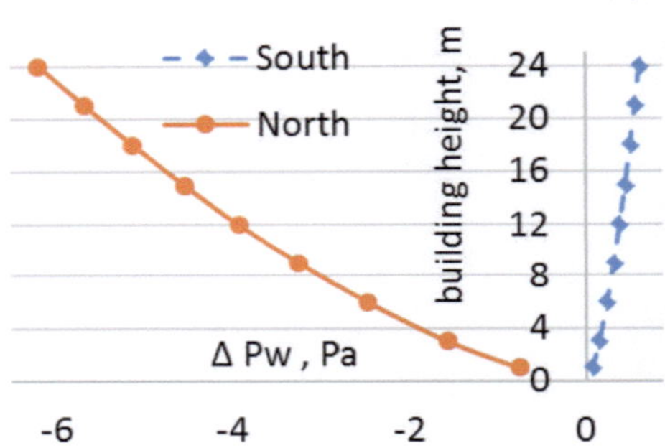

Fig. 4.13 Wind effect pressure change

At the SE wind direction for rooms of the SE orientation, C_p (θ) is positive, for the N—negative.

Figures 4.12 and 4.13 show that the air exchange rate is more significantly affected by the pressure difference caused by the stack effect than by the wind effect.

Figure 4.14 shows the average change in air exchange rate by height for N and SE orientation rooms. The difference between the profiles of change in air exchange rate by height for N and SE orientation is explained by the predominant SEE wind direction.

Figure 4.15 shows the hourly variation of the air exchange rate for representative rooms of N and SE orientation. For the considered input parameters, the air exchange rate fluctuated over three days in the range of 0.53–0.73 h^{-1}. The experimental determination of the air exchange rate (Fig. 4.15b) showed that the air exchange rate for the premises of the 1st floor is at the level of 0.6 h^{-1}. Simulation modelling describes the experimental results with sufficient accuracy [8].

For modern double-glazed windows, an hourly calculation of the natural component of the air exchange rate at different locations of representative rooms showed that the value does not exceed 0.25 h^{-1}, triple glazing in wooden double glazing—0.4 h^{-1}, double glazing—0.65 h^{-1}. It should be noted that on the ground floors the

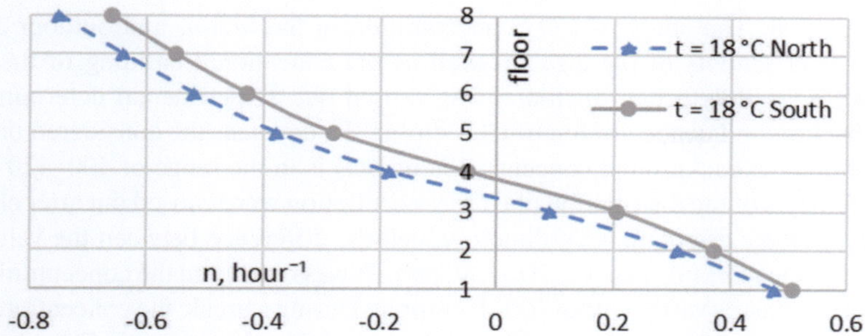

Fig. 4.14 Average change in air exchange rate by building height for rooms with SE and N orientation

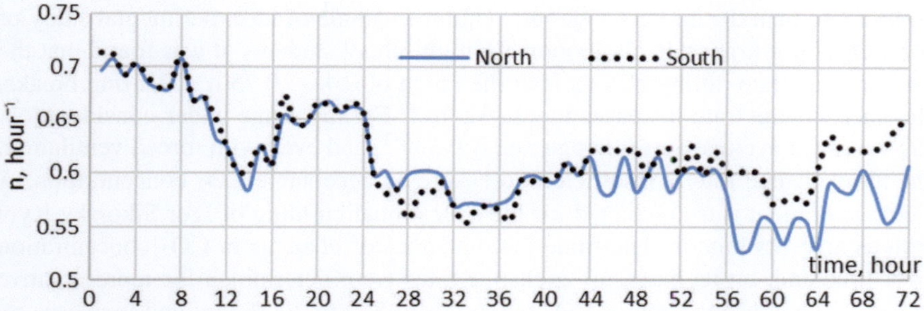

Fig. 4.15 Hourly change in air exchange rate for the ground floor rooms of the SE and NW orientation

natural component of the air exchange rate has a positive value (air infiltration), while on the upper floors it has a negative value (air exfiltration), which should be taken into account when dynamically modelling the energy performance of a building. To maintain the normative value of the air exchange rate when people are in the room, it is necessary to additionally use mechanical ventilation or forced ventilation of the premises.

It has been established that the maximum estimate of additional costs for energy consumption due to natural ventilation (infiltration) compared to the energy consumption of a building with only transmission losses ($n = 0$) differs by 23% for northern and 43% for southern orientations. It has been established that the energy demand of a building at the standard value of the air exchange rate ($n = 1\ h^{-1}$) and the calculated value of the natural component of the air exchange rate differs by up to 50–75%, which is the maximum possible level of savings under actual air exchange conditions compared to the standard. These savings can be reduced by increasing air exchange during occupancy hours, for example, through ventilation.

The chapter presents an integrated approach to the determination and analysis of data on the concentration of CO_2 in the premises and the rate of air exchange in educational institutions under natural air exchange and in the absence of mechanical air circulation. The study of CO_2 concentration in the indoor and outdoor air of three typical schools of the 80 s, as well as the educational building of the Igor Sikorsky Kyiv Polytechnic Institute, was carried out. Experimental determination of the background concentration of CO_2 during the day near the considered objects showed that the background concentration of CO_2 is in the range of 400–420 ppm. Measurements of the distribution of CO_2 concentration were carried out after classes throughout the classroom, according to which the difference between the values at the level of the working area was 30…180 ppm. It was found that the concentration of CO_2 during classes varies within 700–1100 ppm. During a break, the concentration of CO_2 decreases to 500–1000 ppm depending on the type of ventilation. Experimental data on the dynamics of changes in the concentration of CO_2 in the premises are used to determine the air exchange rate based on the balances of air flows and CO_2. It is shown that the number of people present affects the CO_2 concentration in the room more than the air exchange rate. On the example of an experimental study of the CO_2 concentration in classrooms for high school students, it was found that the air exchange rate during classes is in the range of 0.4 … 0.75 h^{-1}. During breaks, the air exchange rate increases to 2.9–3.5 h^{-1}. For the range under consideration, the weighted average air exchange rate is 0.8 h^{-1}, and even with forced ventilation, the air exchange rate is insufficient to ensure an acceptable CO_2 concentration. A field experiment was conducted for the educational building of Igor Sikorsky Kyiv Polytechnic Institute to determine the dynamics of changes in CO_2 concentration over time and, on its basis, air exchange rates were determined for representative classes. The concentration of CO_2 ranged from 500 to 2000 ppm and increases by 350–850 ppm depending on the use and location of classrooms. According to the experimental data, the air exchange rate for the educational building of the school is in the range of 0.35–0.7 h^{-1}. During periods of ventilation, air exchange can increase by 0.45 h^{-1}, but this does not allow to achieve the standard value. In analysing the results, we used simulation models of natural air exchange in the studied audiences based on the improved ASHRAE method. The natural air exchange rate based on modelling is in the range of −0.8…0.5 h^{-1}. Negative values are explained by exfiltration, which is typical for the upper floors. Not only the comfort and condition of the building envelope, but also the overall energy consumption of the building depend on the actual level of air exchange. The ventilation component accounts for 30–60% of the overall energy balance [1].

4.3 Experimental Studies of Air Exchange Rates in the Old Housing Stock

The objects of the study are apartments in multi-apartment residential buildings that correspond to the characteristics of mass construction in the 1980s and a modern residential building with increased requirements for thermal protection built in 2016.

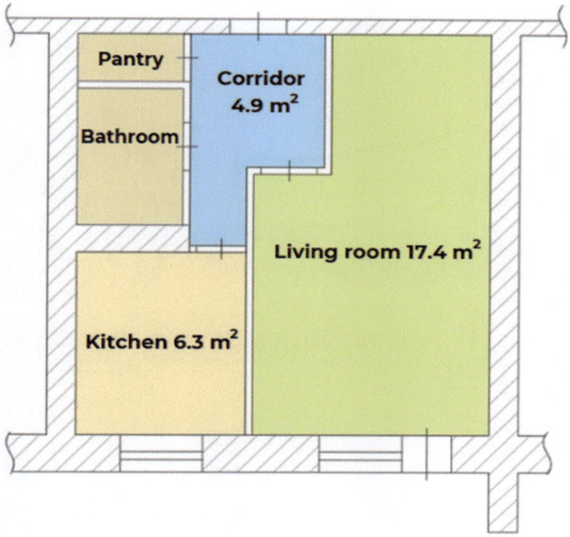

Fig. 4.16 Plan of the research object (old housing stock)

An existing one-room apartment in the family-type dormitory No. 22 of the Igor Sikorsky Kyiv Polytechnic Institute in Kyiv was chosen as the object of studying the impact of natural air exchange on energy consumption. The apartment is located on the ninth floor of a nine-storey building. The apartment has one outer wall facing south. The residential building has a technical floor. The total floor area of the apartment, without a balcony, is 32.8 m^2, the height of the premises is 2.5 m. The basic geometric parameters of the apartment model under study are shown in the plan in Fig. 4.16.

The apartment is divided into five zones corresponding to the rooms: a living room, a corridor, a kitchen, a bathroom and a storage room.

The first step in conducting experimental studies in a residential apartment was to measure changes in CO_2 concentration in the rooms where residents are most often present under normal/actual operating conditions. The measurements were carried out over 10 days in April. A typical change in CO_2 concentration in the living room and kitchen during the day is shown in Fig. 4.17.

The study found that fluctuations in indoor carbon dioxide levels depended on a number of factors. These included the number of people in the room, the duration of their stay, the airing schedule, and climatic changes in the environment. At night, when the residents were sleeping, there were no sharp drops in the concentration of CO_2 in the premises. During the day, on the contrary, people's activity had a significant impact on the level of CO_2. Sharp jumps in the graphs are usually explained by periods of airing the apartment. The average level of carbon dioxide in the apartment during the day was kept at 1200–1500 ppm.

In order to obtain more accurate calculation results, it was decided not to take into account the CO_2 intake from people, so further series of measurements were

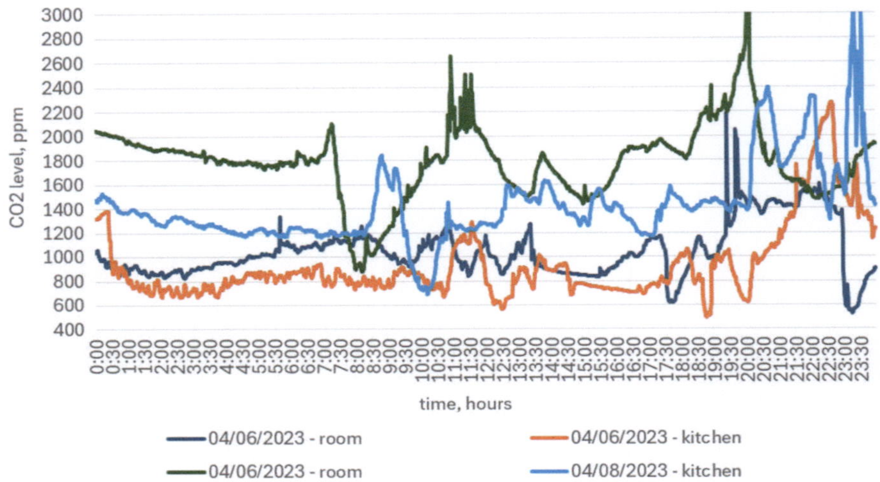

Fig. 4.17 Changes in carbon dioxide concentration during the day in the apartment under actual conditions

carried out in the absence of people in the apartment. In addition, the sensitivity of the devices did not allow us to immediately determine with sufficient accuracy from the balances of changes in CO_2 concentration the level of air transfer between the rooms of the apartment under actual operating conditions, when there were slight fluctuations in CO_2. Therefore, during the experimental measurements, the level of CO_2 in the premises was artificially increased by burning dry alcohol, so that changes in carbon dioxide concentration could be recorded for a longer period of time and a more qualitative analysis of the results could be carried out.

As part of the following series of studies, experimental measurements of carbon dioxide concentration were simultaneously carried out in the living room, kitchen, corridor of the apartment and in the ventilation duct leading to the kitchen with a 1-min step. The plan of the experiment is shown in Fig. 4.18.

As part of the study, experimental measurements of carbon dioxide concentrations in the apartment and classroom were made in 1-min increments.

The plan of the experiment is shown in Fig. 4.18. The climatic data was also recorded, namely the temperature in the apartment, the outdoor temperature, wind direction and strength.

Measuring instruments. Three types of gas analysers were used for the experimental measurements: a TR-75Ui CO recorder_2 (Fig. 4.19a), an air quality analyser D91 (Fig. 4.19b), and a Xintest HT-501 carbon dioxide detector-datalogger with a thermohygrometer (Fig. 4.19c).

The complex device "TR-75Ui" has a range of CO_2 measurements: 0–9999 ppm (±40 ppm ± 3%); temperature: 0–55 °C; relative humidity: 10–95%.

The D91 air quality analyser is equipped with various sensors for the simultaneous evaluation of carbon dioxide concentrations: 400–4000 ppm (±70 ppm ± 3%);

Experiment plan

1. Check the devices for serviceability and charge level
2. Switch on the devices and place them at representative points under actual operating conditions
3. Switch on the devices and place them at representative points under artificially created conditions
4. Close all windows, entrance and interior doors
5. In the selected room, place dry alcohol in a prepared vessel and light it

Fig. 4.18 Plan of the experiment

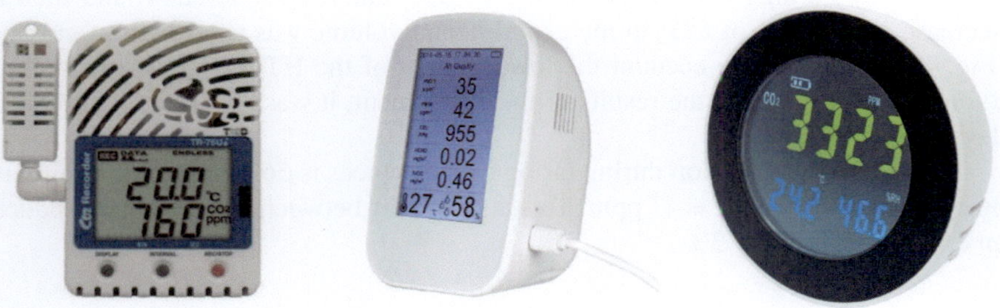

a) b) c)

Fig. 4.19 Devices for measuring CO_2 concentration in an apartment [9]: **a** complex device for CO registrar$_2$ “TR-75Ui”; **b** air quality analyser “D91”; **c** detector-alarm “HT-501”

volatile organic compounds (TVOC: benzene, styrene, phenols, toluene, xylene, etc.:) 0–2.0 mg/ m^3; formaldehyde (HCOH): 0–2.5 mg/m^3; microparticles (dust, smoke): 0–6 mg/ m^3; temperature: 0–55 °C; humidity: 0–100%.

Detector-alarm “HT-501” is a portable CO gas analyser$_2$ with a built-in thermo-hygrometer designed to monitor indoor microclimate parameters. The built-in data logger allows you to transfer data to a laptop: store a numerical series and build graphs. The carbon dioxide measuring range of the HT-501 is: 0–9999 ppm (±40 ppm ± 3%); temperature: (−10)-(+70 °C) (±0.3°C (10 to 50°C), ± 1.2°C (other range)); humidity: 0.1–99.9% (± 3%).

As part of the study, we checked the uniformity of the distribution of carbon dioxide concentration in different places of each of the studied premises. To do this, five carbon dioxide measuring devices were installed in each room. These devices were located at a height of 1–1.5 m from the floor level and in different parts of the room, which allowed us to get a more detailed picture of the CO distribution 0.2

The analysis of the readings from all five devices showed that the difference between them is within the error margin. This indicates that the distribution of carbon dioxide in the room is uniform. Thus, to monitor the carbon dioxide level, it is enough to use only one device in each room at the same height. This greatly simplifies the monitoring process and reduces equipment costs.

During a series of experimental studies, the carbon dioxide concentration and temperature (both indoor and outdoor) were measured simultaneously in the living room, apartment corridor, kitchen and in the ventilation duct leading to the kitchen with a 1-min step. The values of the change in the background (external) CO $level_2$ and in the corridor of the building were also recorded. During the research, the devices were placed at a height of 1 m.

This approach allows us to take into account the movement of air masses between rooms. The data obtained was used to calculate the air exchange rate and analyse the indoor air quality.

Convergence of the results of the devices. Measuring devices were placed in a 0.3 m^3 (cubic) test volume with a small slit. The level of carbon dioxide was artificially increased by burning dry alcohol in the volume under consideration, and then a decrease in the level of CO_2 in the experimental volume was recorded for an hour (Fig. 4.20). Taking into account the lower inertia of the HT-501 device according to the passport data and the results of the experiment, it was chosen as a reference device.

The maximum deviation during taring of the devices is 56 ppm, the minimum is 36 ppm, and the average is 47 ppm. This discrepancy between devices can introduce an error not exceeding 3%.

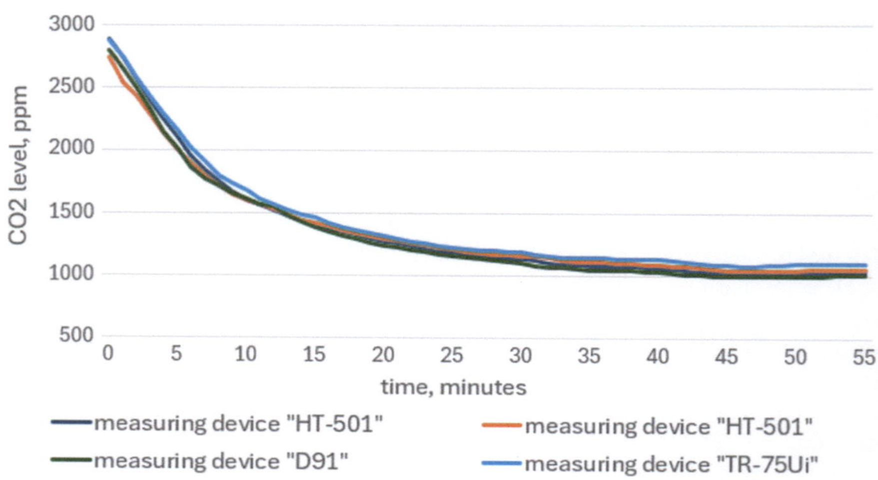

Fig. 4.20 Convergence of device results

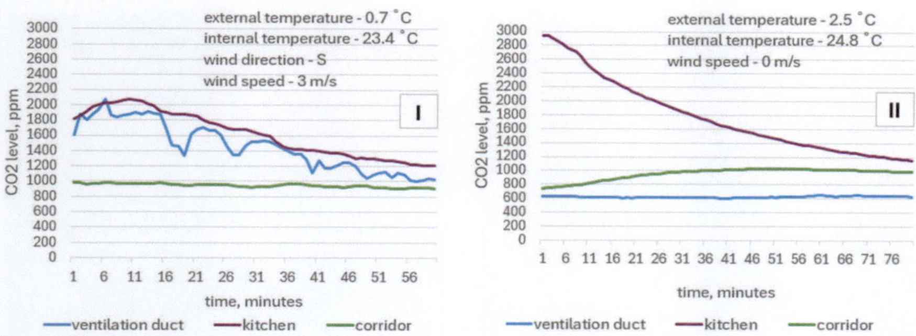

Fig. 4.21 Changes in carbon dioxide concentration in the apartment premises with increasing CO_2 levels in the kitchen under different climatic conditions

The temperature inside the apartment was also measured. Actual climatic data, such as outdoor temperature, wind direction and strength, were collected from the weather station closest to the study site [10].

The next series of studies of changes in the concentration of carbon dioxide in the apartment was carried out in winter and summer. The series of measurements during the heating period included 13 measurement intervals of 60–120 min over 10 days.

The southeast wind direction in February is typical for the city of Kyiv, respectively [11]. During the study period, according to the actual climatic data, the southern and southwestern wind directions were typical. The average daily outdoor air temperature was (+4.1 °C)-(+6 °C), with daily fluctuations of (+6 °C), (−3 °C). Also, at the beginning and end of the experiment, the CO_2 level in the dormitory corridor was recorded, which was within 540–570 ppm. The level of carbon dioxide in the outside air during the experiments was in the range of 390–430 ppm [12].

In order to determine the mechanism of air mass transfer between the rooms of the apartment, an artificial increase in the level of CO_2 was carried out in different rooms. Figure 4.21 shows the results of measurements under conditions of increasing the level of CO_2 in the kitchen.

Analysing the typical graphs in Fig. 4.21, we can conclude that the air movement in the ventilation duct depends on the wind direction. With a south and south-west wind direction, the air from the kitchen room flowed into the ventilation duct—movement mechanism I (Fig. 4.21a), because the change in carbon dioxide concentration in the kitchen room and the ventilation duct has similar trends. In the case of a westerly wind direction and in the absence of wind (Fig. 4.21b), on the contrary, the supply air entered the apartment through the ventilation duct—movement mechanism II, because the level of carbon dioxide in the ventilation duct remained relatively constant throughout the experiment at the level of the external environment [12].

As a result of the research, two directions of air exchange in the apartment were identified, namely infiltration and exfiltration. In the first case, the air came in through the windows and the front door of the apartment and was exhausted through the ventilation duct (Fig. 4.22a). In the second case, the supply air came in from the

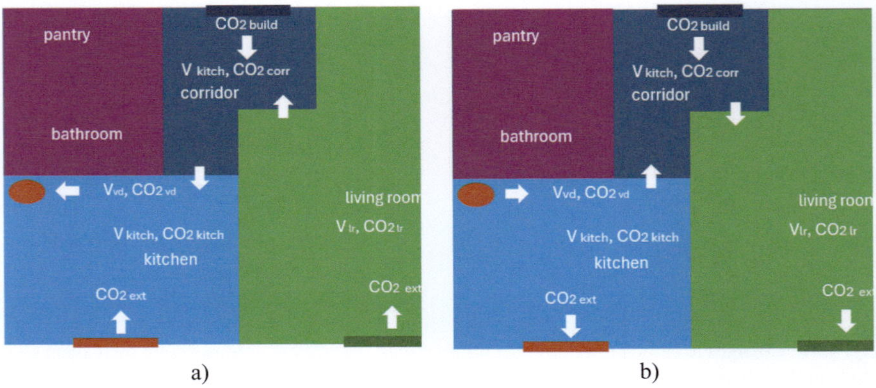

Fig. 4.22 Mechanisms of air exchange in an apartment [12]: **a** supply air enters through windows and the front door; **b** supply air enters through the ventilation duct and the front door: V_{lr}, V_{corr}, V_{kitch}, V_{vd}—is the volume of the premises of the room, corridor, kitchen and ventilation duct, respectively, m^3; C_{lr}, C_{corr}, C_{kitch}, C_{vd}, C_{ext}, C_{build}—CO concentration$_2$ in the premises of the living room, corridor, kitchen, ventilation duct, background and in the corridor of the building, respectively

ventilation and through the front door, and was exhausted through the windows in the kitchen and room (Fig. 4.22b).

Below, as an example, is a system of equations for calculating the internal CO_2 concentration at constant background carbon dioxide concentrations outside the apartment, on the street and in the corridor of the building (common areas) for the case when the supply air was supplied by infiltration (system of Eq. 4.2) and exfiltration (system of Eq. 4.3):

$$\begin{cases} CO_{2lr} = CO_{2lr0} + \frac{V_{supp.lr}}{V_{lr}} \int (CO_{2ext} - CO_{2lri})_\tau d\tau \\ CO_{2corr} = CO_{2corr0} + \frac{V_{supp.lr}}{V_{corr}} \int (CO_{2lri} - CO_{2corri})_\tau d\tau \\ \quad + \frac{V_{supp.build}}{V_{corr}} \int (CO_{2build} - CO_{2corri})_\tau d\tau \\ CO_{2kitch} = CO_{2kitch0} + \frac{V_{supp.corr}}{V_{kitch}} \int (CO_{2corri} - CO_{2kitchi})_\tau d\tau \\ \quad + \frac{V_{supp.kitch}}{V_{kitch}} \int (CO_{2ext} - CO_{2kitchi})_\tau d\tau \\ V_{supp.corr} = V_{supp.lr} + V_{supp.build} \end{cases}, \tag{4.2}$$

$$\begin{cases} CO_{2lr} = CO_{2lr0} + \frac{V_{supp.corr}}{V_{lr}} \int \left(CO_{2supp.corr} - CO_{2lri}\right)_\tau d\tau \\ CO_{2corr} = CO_{2corr0} + \frac{V_{supp.kitch}}{V_{corr}} \int (CO_{2kitchi} - CO_{2corri})_\tau d\tau \\ \quad + \frac{V_{supp.build}}{V_{corr}} \int (CO_{2build} - CO_{2corri})_\tau d\tau \\ CO_{2kitch} = CO_{2kitch0} + \frac{V_{supp.vd}}{V_{kitch}} \int (CO_{2vdi} - CO_{2kitchi})_\tau d\tau \\ \quad + \frac{V_{supp.kitch}}{V_{kitch}} \int (CO_{2ext} - CO_{2kitchi})_\tau d\tau \\ V_{supp.\text{kop}} = V_{supp.kitch} + V_{supp.build} \end{cases} \tag{4.3}$$

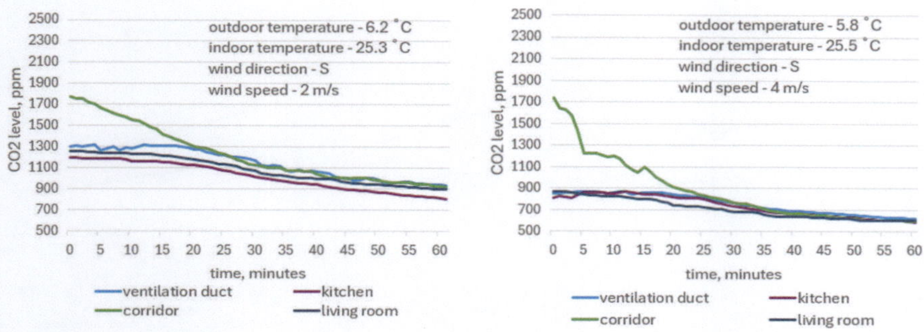

Fig. 4.23 Changes in carbon dioxide concentration in the apartment premises when injecting CO_2 into the kitchen under different climatic conditions [12]

where are the input parameters: CO_{2lr0}, CO_{2corr0}, $CO_{2kitch0}$—CO_2 concentrations at the current time in the room, corridor, kitchen and ventilation duct, respectively, g/m^3; CO_{2lri}, CO_{2corri}, $CO_{2kitchi}$, CO_{2vdi}—CO_2 concentrations at the current time in the premises of the room, corridor and kitchen, respectively, g/m^3; V_{lr}, V_{corr}, V_{kitch}—volume of the premises of the room, corridor and kitchen, respectively, m^3; CO_{2lr}, CO_{2corr}, CO_{2kitch}, CO_{2vd}, CO_{2ext}, CO_{2build}—CO_2 concentrations in the premises of the room, corridor, kitchen, ventilation duct, background and corridor of the building, respectively, g/m^3.

Determined: $V_{supp.lr}$, $V_{supp.corr}$, $V_{supp.kitch}$, $V_{supp.build}$, $V_{supp.vd}$—volume of fresh air supply to the room, corridor, kitchen, from the ventilation duct and corridor of the house, respectively, m^3/h.

The system of equations was solved using the Mathcad software, taking into account the dynamics of input parameters over time.

Figure 4.23 shows the results of measurements of carbon dioxide levels in the kitchen, corridor, living room and ventilation duct under conditions of increasing CO_2 levels in the corridor.

The data shown in Fig. 4.23 indicate that the CO_2 concentration in the ventilation duct is significantly higher than the background CO_2 concentration in the ambient air. The graph of CO_2 concentration in the ventilation duct is similar to the graph of CO_2 concentration in the kitchen, indicating that the air from the kitchen is exhausted through the ventilation duct. Based on several series of measurements, it was found that the ventilation in the apartment is largely dependent on the natural ventilation flow created through the windows.

The graphs (Fig. 4.24) show the results of typical experiments where the starting concentration in different rooms of the apartment was different under relatively similar indoor and outdoor conditions. At the same time, the determined air exchange rate remained at approximately the same level: 0.74 h^{-1} for the living room, 0.66–0.81 h^{-1} for the corridor, 3.4 h^{-1} for the kitchen. This means that the air exchange rate was approximately the same regardless of the initial CO_2 concentrations.

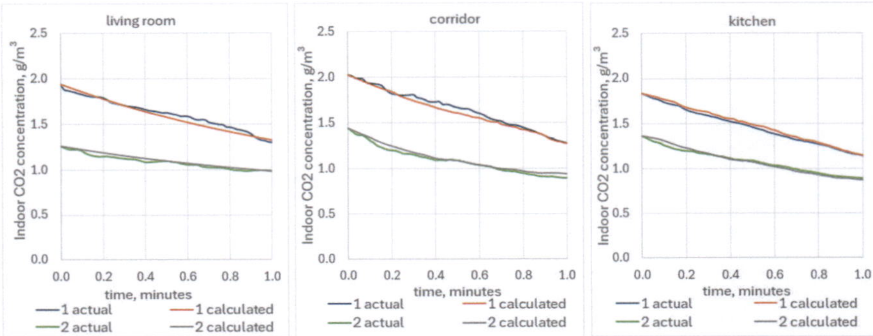

Fig. 4.24 Changes in the concentration of carbon dioxide in the apartment under conditions when the supply air comes through the windows and the front door (indices 1, 2 correspond to the number of the experiment) [12]

In classical calculations [13], it is assumed that fresh air enters rooms through infiltration, namely through leaks in windows and doors, and exits through the ventilation duct. The results of the study indicate that there are two directions of air movement. Figure 4.25 shows the results of processing a typical study conducted under conditions where the supply air enters through the ventilation duct and the front door. Given that the direction of air movement in this experiment corresponds to Fig. 4.26a, there are differences in the values of the air exchange rate compared to the direction of air movement shown in Fig. 4.26b. For the above conditions, the air exchange rate for the living room was 0.37 h^{-1}, for the corridor—1.06 h^{-1}, and for the kitchen—0.57 h^{-1}. The change in air exchange rate is explained by a different difference in air density and temperature.

A series of experimental studies of changes in CO_2 concentration in representative rooms of an apartment in the summer were also conducted. This series of measurements included 18 measurement periods of 60–120 min during the month of August.

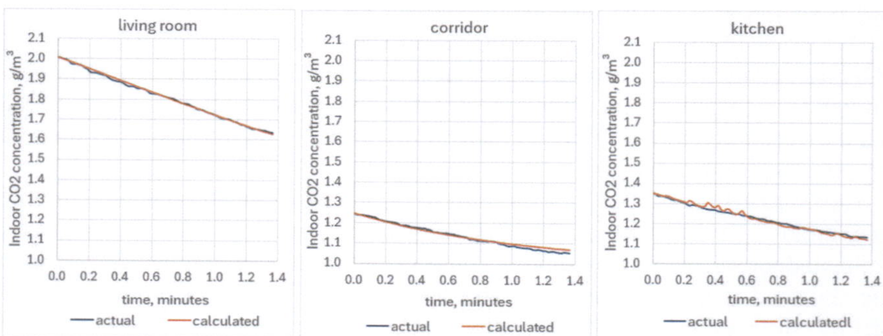

Fig. 4.25 Changes in carbon dioxide concentration in the apartment premises when the supply air comes through the ventilation duct and the front door [12]

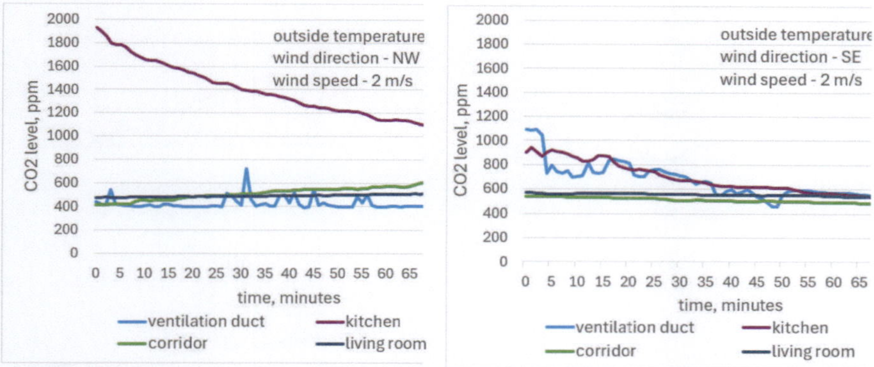

a) b)

Fig. 4.26 Changes in carbon dioxide concentration in the apartment [12] for infiltration (**a**) and exfiltration (**b**) mechanisms

The measurements were carried out according to the experimental plan shown in Fig. 4.18. For the summer period in the city of Kyiv, westerly, northwesterly, and northerly wind directions are typical [11]. During the study period, the prevailing wind direction was northwest. Measurements were taken during the daytime, when the outside temperature was (+21 °C)-(+30 °C). Also, at the beginning and end of the experiment, the CO_2 level in the dormitory corridor was recorded, which was within 410–460 ppm. The level of carbon dioxide in the outside air during the experiments was in the range of 410–435 ppm.

The measurement results clearly show the dependence of infiltration and exfiltration on the wind direction. With a north-westerly wind direction, fresh air entered through the ventilation duct (Fig. 4.26a). In the south, southwest, and southeast wind directions, air entered through windows and doors and exited through the ventilation duct (Fig. 4.26b), as the trends in carbon dioxide concentration in the kitchen and ventilation duct are similar.

After processing all the series of experiments, it was found that the derived systems of Eqs. (4.2) and (4.3) describe the actual changes in CO_2 in the premises for both air transfer mechanisms quite accurately. Figure 4.27 shows graphs of the calculation results of typical experiments under conditions of increasing carbon dioxide levels in different rooms of an apartment. The maximum discrepancy between the actual and calculated values was ±50 ppm and is associated with the error of measuring devices, as well as with the differences in CO_2 concentration in the ventilation duct, which is common to the entire riser.

In winter, fresh air was sucked into the apartment through the windows, while in summer, infiltration and exfiltration were recorded in almost equal amounts. The actual air exchange rate for both mechanisms of air movement in the apartment was also determined. The windows and balcony doors of the apartment are oriented to the

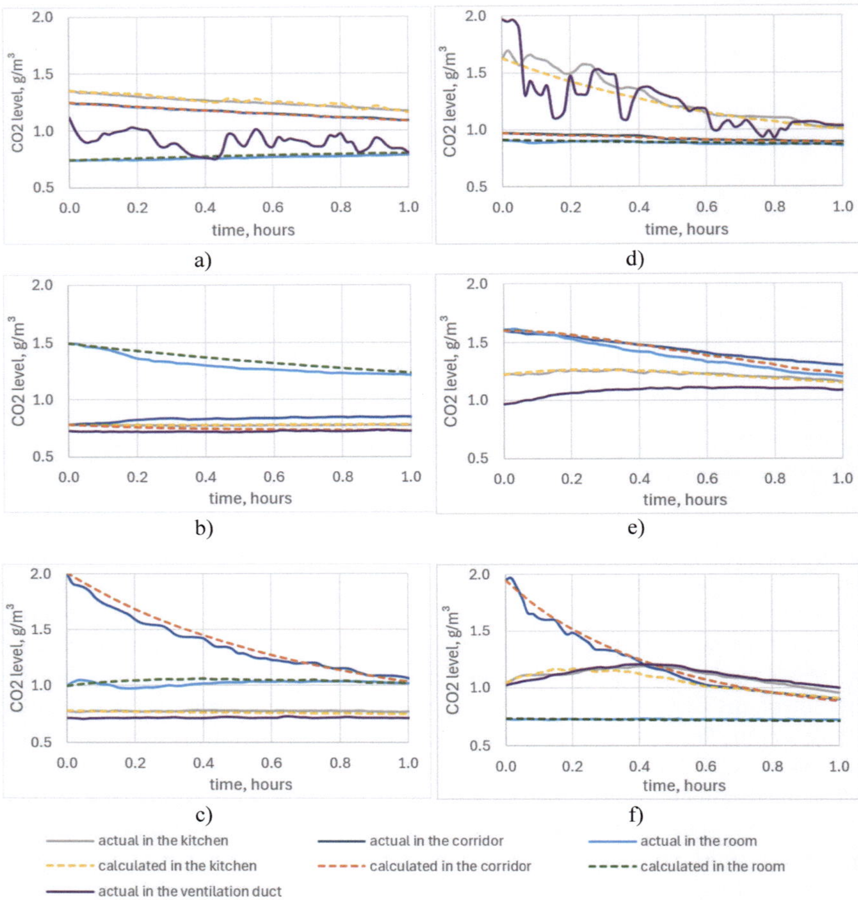

Fig. 4.27 Actual and calculated concentrations of carbon dioxide in the apartment premises under conditions of exfiltration (**a–c**) and infiltration (**d–f**) with increasing levels of carbon dioxide in different rooms of a representative apartment **a**, **d**s—increase in CO_2 in the kitchen; **b**, **d**—increase in CO_2 in the room; **c**, **e**—increase in CO_2 in the corridor [14]

south, respectively, with south, southeast and southwest wind directions, the value of the air exchange rate is higher.

In the summer, the air exchange rate for the infiltration mechanism was 0.57 h^{-1} for the living room, 2.34 h^{-1} for the corridor and 2.41 h^{-1} for the kitchen. For the exfiltration air movement, the average air exchange rate was 0.24 h^{-1} for the living room, 0.94 h^{-1} for the corridor and 0.52 h^{-1} for the kitchen.

Based on experimental studies, it was found that the actual air exchange rate when fresh air enters the room through windows and doors exceeds the minimum standard values. However, when the air flows in the opposite direction, this indicator does not meet the regulatory requirements for the level of air exchange. Taking into account the direction of movement of the air entering the apartment, in fact, the energy consumed for heating the supply air will be proportional to the air exchange rate of the air coming from outside.

4.4 Experimental Studies of Air Exchange Rates in Modern Housing Stock

The object of the study is an existing residential premises. This is a two-room apartment located on the fourth floor of a five-storey residential building in Kyiv, built in 2016. The apartment under study has two exterior walls. The total floor area of the apartment, without a balcony, is 49.44 m^2, the height of the premises is 2.7 m. The basic geometric parameters of the apartment model under study are shown in the plan in Fig. 4.28. When creating the energy model, the apartment was divided into five zones corresponding to the rooms: two bedrooms, a kitchen, a bathroom, and a corridor. This decision was made to take into account the fact that each room in the apartment has specific behavioural and operational characteristics. The data on the area and heat transfer resistance of the building envelope of the study object are summarised in Table 4.1 [15].

The heat supply source for the apartment is an individual double-circuit gas boiler, which is used to provide heating and hot water supply circuits. The apartment is heated for 176 days of the heating period. The internal water heating system of the apartment is two-pipe, horizontal. Heating appliances (radiators) in the bedrooms

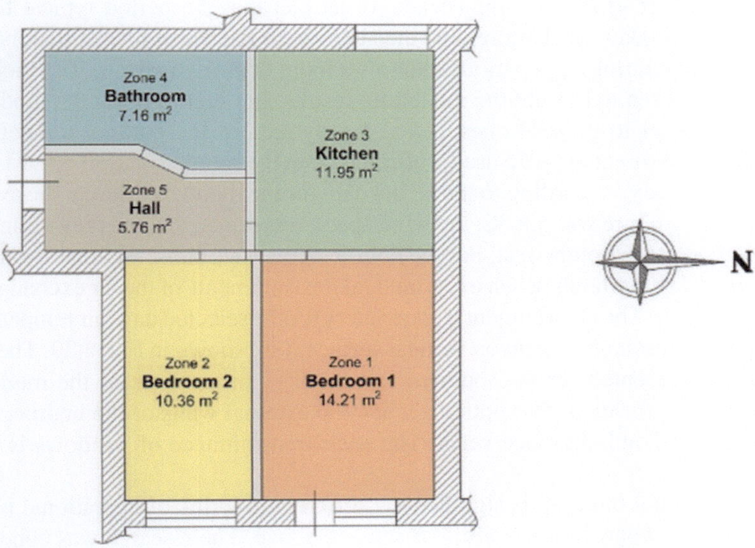

Fig. 4.28 Plan of the research object (modern housing stock)

Table 4.1 Characteristics of enclosing structures [15]

Type of construction	Heat transfer resistance of the structure[4.1], R_i, $m^2 \cdot K/W$	Construction area, A_i, m^2
External walls	1.506	16.1
Windows and balcony doors	0.945	7.47
Floor	–	49.44
Ceiling	–	49.44

[4.1]heat transfer resistance of the external walls was determined with consideration of heat-conducting inclusions

and kitchen are installed near the external walls under windows without radiation protection, and the bathroom is heated by underfloor heating. The software model stipulates that the air temperature in the rooms is regulated by the residents using an electronic thermostat. The thermostat is equipped with temperature sensors placed in rooms with heating. The presence of the thermostat prevents the temperature from falling below the programmed temperature in the rooms where the sensors are located. The temperature mode is programmed separately for weekdays and weekends [15].

The apartment's ventilation system is naturally induced. Fresh air is brought inside by infiltration (through gaps in windows and doors) and through the window micro-ventilation system. Air exchange occurs due to the pressure difference inside and outside the building. The building is equipped with a system of exhaust ventilation ducts, which are equipped with mechanical exhaust systems at the exit points. The capacity of the kitchen hood is 400–800 m^3 /h, depending on the selected mode of operation. The capacity of the exhaust fan in the bathroom is 125 m^3 /h [15].

The impact of global warming leads to the fact that the period typical for the off-season now takes much more time than before, so the use of intermittent heating and ventilation during such periods requires a more detailed analysis. The month of March was chosen to present the modelling results. In order to bring the modelling conditions closer to the real ones, actual hourly data of the outdoor temperature, wind speed and direction were used, collected from the weather station closest to the object under study. According to the actual data for the month of March, the average monthly temperature was 5.6 °C, the wind speed was 4 m/s, and the prevailing wind direction was south-south-west. For the hourly studies, we chose weather conditions for a typical day in March, when experimental measurements of the air exchange rate were carried out. The environmental parameters for the selected day (air temperature, wind speed and solar heat gain on vertical surfaces) are shown in Fig. 4.29. The solar heat gain is presented for the southern and northern orientation, as the modelling was carried out for these two options, when the external walls of the bedrooms are oriented to the south and vice versa. The solar transmittance of windows is 0.579 [15].

The residential building is significantly different in terms of operational performance from the educational facilities discussed above. The research was conducted in a 2-room apartment that meets modern thermal and physical requirements, located

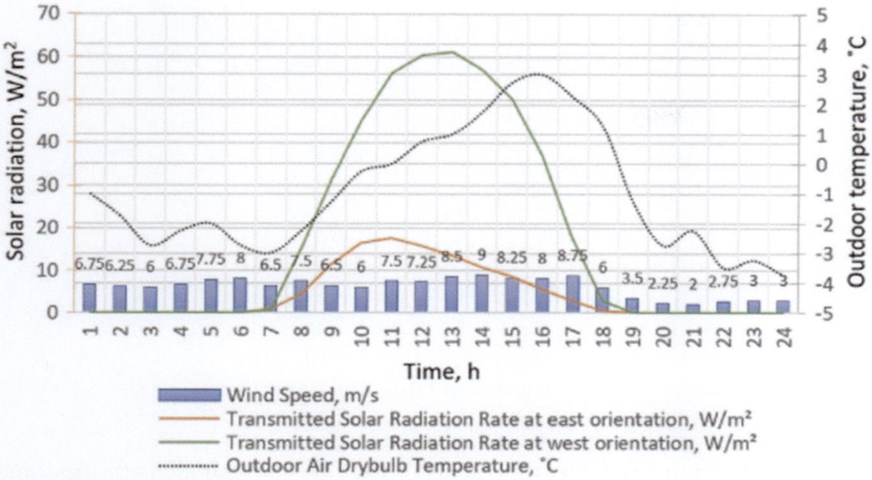

Fig. 4.29 Environmental parameters for a representative day

on the 4th floor of a 5-storey building. Measurements of CO_2 concentration were recorded in 1-min increments. The following full-scale experiments were conducted to measure the CO_2 concentration in the rooms and kitchen: (1) with and without additional internal CO_2 emissions; (2) with and without micro-ventilation in the rooms. The building meets modern thermal insulation requirements, and therefore, for this type of building, the ventilation component is the largest in the overall energy balance of building losses [13].

Given that CO_2 emissions in residential premises are lower than in educational institutions. Prior to the experiment, the concentration of CO_2 was artificially increased by burning dry alcohol tablets in the studied premises. This will allow us to better monitor the dynamics of changes in the concentration of CO_2 in the premises.

At the beginning of the study, when the window and door in room 1 are closed, the concentration of CO_2 is 2000 ppm, after an hour with the door to the corridor open, the concentration of CO_2 decreased by 1700 ppm, i.e. by 300 ppm. When the door and window in room 1 are closed, a similar change in CO_2 concentration by 300 ppm occurs in 3 h. If the window is open for micro-ventilation and the door is closed, the CO_2 concentration in room 1 changes from 3000 to 580 ppm in 2 h [13].

In room 2, with the windows and doors closed and no people present, the CO_2 concentration changes from 1950 to 1250 ppm in 6 h. With the window open for micro-ventilation and the internal interior door open, the CO_2 concentration decreases from 2850 to 1050 ppm, i.e. by 1800 ppm, in 2 h. The concentration of CO_2 in room 2 decreases more slowly than in room 1, which is explained by the fact that the window from room 2 opens onto a glazed balcony, unlike in room 1. It should be noted that the high concentrations of CO_2 in the room were created artificially to study the dynamics of CO_2 changes by different ventilation mechanisms [13].

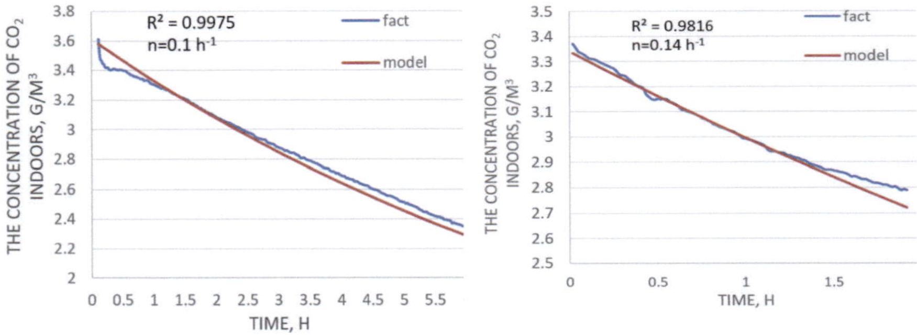

a) b)

Fig. 4.30 Air exchange in room 1 (**a**) and room 2 (**b**) without ventilation

The kitchen has a 200 m3/h extractor hood. The CO_2 concentration was measured with and without the hood in operation. During cooking on a gas stove for 30 min, the CO_2 concentration increases from 1500 to 2200 ppm, i.e. by 700 ppm without the hood operating and the window closed. Under the condition of intermittent operation of the hood during the cooking period, the concentration of CO_2 had an increasing and decreasing trend and was in the range of 2200…1300 ppm. At first, a decrease of 910 ppm in 1 h was observed, followed by an increase of 300 ppm in 15 min (the hood was switched off), and the CO_2 concentration decreased by 200 ppm for the next 30 min [13].

Using formula (3.14), the experimental data were processed to determine the actual air exchange rate. The air exchange rate with closed windows and doors is 0.14 h in room 1^{-1}, 0.1 h in room 2^{-1}, and 0.3 h in the kitchen^{-1} (Fig. 4.30). The analysis of air exchange balances in an apartment shows that air infiltration is typical for living rooms, and exfiltration in the kitchen [13].

In Fig. 4.30 shows the change in CO_2 concentration in residential premises (in room 1—Fig. 4.30a, in room 2—Fig. 4.30b) without ventilation and in the absence of people.

Figure 4.31 shows the change in the CO_2 concentration of in residential premises (room 1—Fig. 4.31a, room 2—Fig. 4.31b) with micro-ventilation. In contrast to Fig. 4.30, in Fig. 4.31, a greater curvature is observed, which is explained by an increase in the amount of fresh air entering the room.

Under the condition of micro-ventilation in the rooms, the following air exchange rate is created in: room 1—1.4 h^{-1}, room 2—0.73 h^{-1} (windows overlook the glazed balcony) (Fig. 4.31), kitchen—1.4…1.5 h^{-1} (hood is working) (Fig. 4.31) [13].

Figure 4.32 shows the change in the CO_2 concentration over time for intermittent hood operation. Figure 4.32 shows that during the period when the hood is not operating and the window is closed, the air exchange rate is 0.15 h^{-1}.

Experimental studies show that in the apartment under consideration, in the absence of artificial ventilation or mechanical ventilation, the air exchange rate is at the level of 0.1…0.155 h^{-1}, which is explained by the high resistance to air

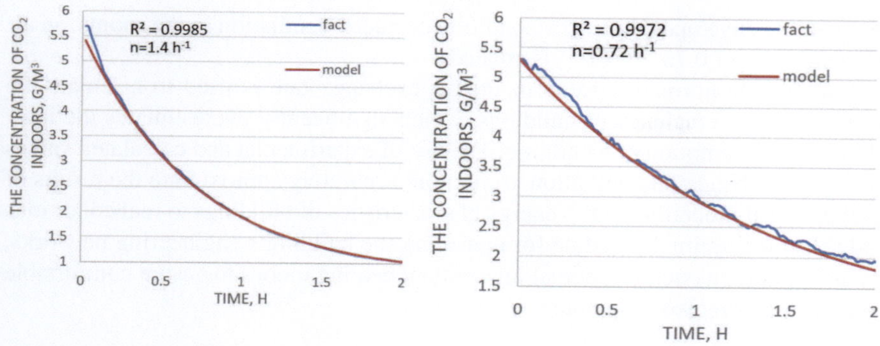

a) b)

Fig. 4.31 Air exchange in room 1 (**a**) and room 2 (**b**) with micro-ventilation

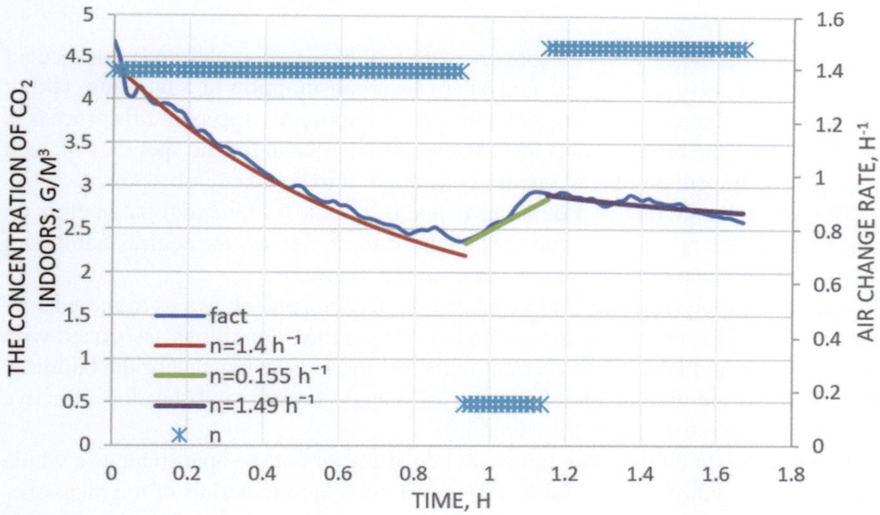

Fig. 4.32 Changes in CO_2 concentration in the kitchen for different operating modes of the exhaust system

permeability of the building envelope. The obtained research results correspond to similar results of other studies [16]. It should be noted that the normative value of the CO_2 concentration level for residential and public buildings is in the range of 750…1000 ppm, the air exchange rate: in residential buildings—0.8 h^{-1}, schools, universities—1 h^{-1}.

Studies conducted in a modern residential building show that in the absence of artificial ventilation or mechanical ventilation, the air exchange rate is at the level of 0.1…0.155 h^{-1}, which is explained by the high resistance to air permeability of the building envelope. Under the condition of micro-ventilation in the rooms, an air exchange rate of 0.73…1.4 h^{-1} is created.

Usually, the normative value of the air exchange rate is used to calculate the level of energy efficiency of buildings, which significantly overestimates the level of energy consumption of a building. The use of experimental and calculated values of the air exchange rate will allow us to more accurately approximate the results of mathematical modelling of the energy characteristics of buildings to real conditions and select the optimal schedule for managing the building's engineering networks, including the selection of optimal intermittent heating modes to ensure comfortable conditions during working hours.

4.5 Analysis of the Impact of Air Exchange Distribution Between Rooms on Energy Consumption

Determining the local energy performance of a building is a complex and multifaceted process that involves a detailed analysis of heat consumption in a building, taking into account influencing parameters. One of the important aspects of this process is the division of the building into local zones, which allows for the specifics of each room and its impact on overall energy consumption to be taken into account.

The air exchange rate in a building is one of the most influential parameters on heat loss and energy consumption, as each air exchange cycle requires additional energy costs to heat or cool the new air entering the room.

A detailed analysis of the air exchange rate in each zone allows us to more accurately estimate the energy consumption in real operating conditions associated with air circulation and take it into account in the overall energy balance of the building. This helps to identify potential sources of energy savings and develop effective strategies to optimise energy consumption.

Below are data on the main dimensional parameters of the apartment as a whole and the main studied rooms (Table 4.2) and construction materials of a typical one-room apartment (Tables 4.3 and 4.4). The representative apartment is described in Sect. 4.4 and corresponds to the characteristics of mass built-up buildings.

When constructing the 3D geometry (Fig. 4.33), the internal volume of the building was divided into zones in order to take into account different air exchange rates in different rooms in the future, using the EnergyPlus software for dynamic

Table 4.2 Key dimensional indicators

	Area, m^2	Height, m	Volume, m^3
Apartment	32.8	2.5	92.5
Living room	17.4	2.5	43.5
Corridor	4.9	2.5	12.3
Kitchen	6.3	2.5	15.8

Table 4.3 Description of the building's structural elements

Structural elements	Description
Overlap	Reinforced concrete slabs
Walls	Brick, plastered, with decorative tiles
Partitions	Brick, plastered
Windows	Single-chamber double-glazed unit in wooden twin glazing
Entrance doors	Metal
Internal doors	Wooden
Balcony doors	Wooden double doors with partial glazing

Table 4.4 Materials and characteristics of enclosing structures

Enclosing structures	Materials	Layer thickness, m
External walls	Plaster	0.010
	Brickwork	0.520
	Facing tiles	0.010
Translucent structures	Glass	0.004
	Air	0.055
	Glass	0.004
Overlap	Reinforced concrete slab	0.250
	Cement screed	0.020
	Parquet	0.020

modelling of building energy performance [17]. The 3D energy model of the building was created in the DesignBuilde software [18].

The apartment has a natural ventilation system. Fresh air enters the apartment through gaps in the building envelope (windows and doors). Air exchange occurs due to the pressure difference inside and outside the building.

The heat gain from people and the lighting system are given according to the graph in Fig. 4.34. The apartment is occupied by 2 people. The internal operating mode of the apartment directly affects the schedule of heat energy use and the internal heat gains from people, lighting and electrical equipment. The basis of the operating mode is the schedule of people's presence. During the hours of maximum load, lighting and electrical equipment in the building is set at 2 W/m^2 [19].

Heat transfer in the building envelope is carried out by two main mechanisms: convection and heat conduction. Given the same temperature conditions in neighbouring apartments, it is assumed that the walls, ceiling and floor bordering other apartments are adiabatic.

In the energy model of the apartment, the main source of heat is district heating. Heating devices (radiators) are installed near the outer walls under the windows in the bedrooms and in the kitchen without radiation protection. There is a heated towel

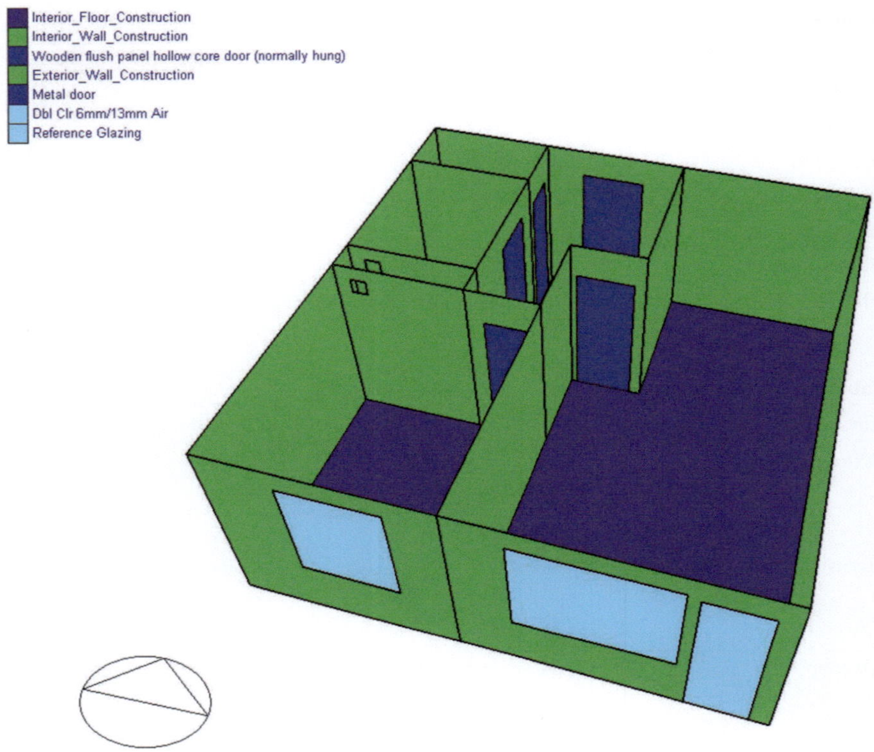

Fig. 4.33 3D model of a building in DesignBuilder

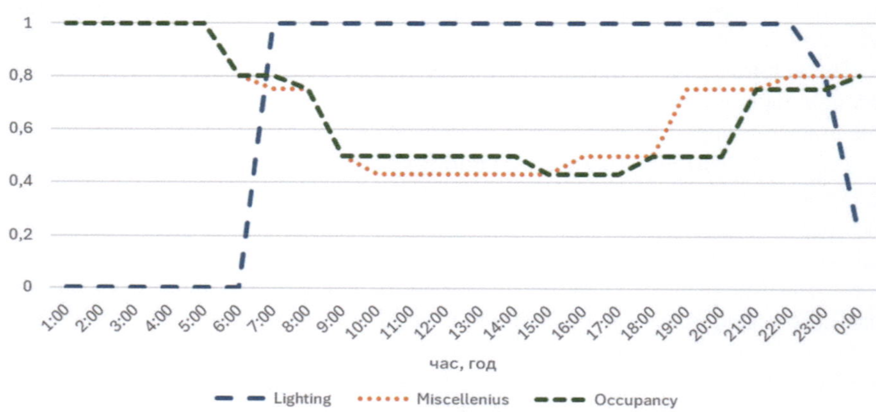

Fig. 4.34 Graph of activity and operation of the lighting system and use of electrical equipment during the day

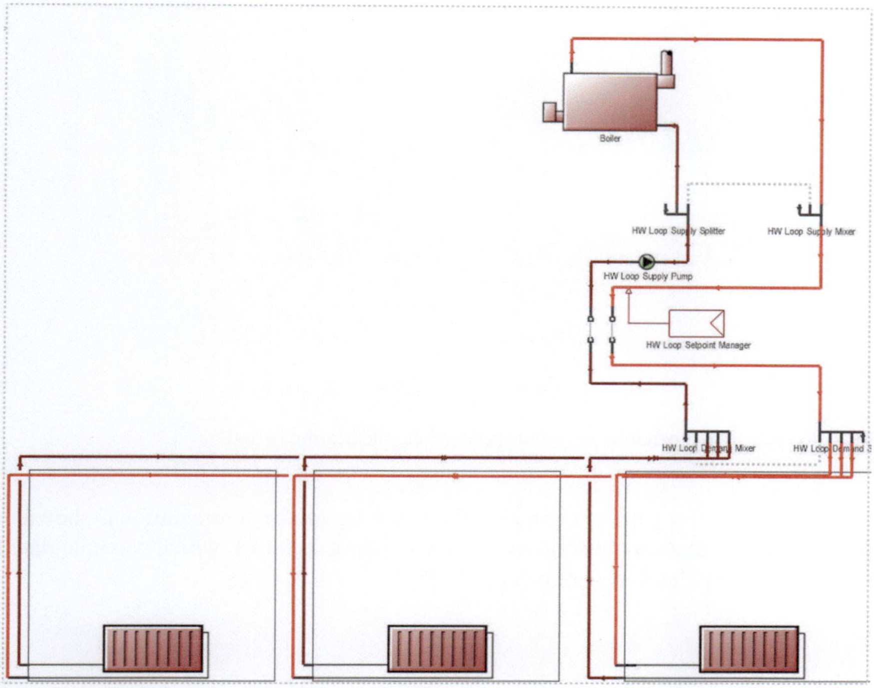

Fig. 4.35 Schematic representation of the apartment heating system

rail in the bathroom. During the heating season, the temperature in the apartment is maintained at 20 °C [19]. The apartment heating system created in the DesignBuilder software environment [19] is shown in Fig. 4.35.

The heat transfer process in an apartment is carried out by radiation and convection from heating appliances. The air in the rooms accumulates part of this heat, while the rest is transferred to the internal and external structures of the building.

The algorithm for calculating heat gain from people used in EnergyPlus [17] takes into account the energy balance, which includes radiant, convective and latent heat of moisture evaporating from the surface of the human body. The software model assumes that the metabolic heat from people living in the apartment is 117 W/person.

The heat gain from the lighting system and electrical equipment was determined in EnergyPlus [17] by the method of setting the load per square metre of floor space, in accordance with the standard [19].

Hourly environmental parameters for the heating period of a typical year were taken from the IWEC international climatological database for the city of Kyiv [20]. Figure 4.36 shows hourly outdoor air temperature and wind speed. In general, for the winter period in Kyiv, inter-day temperature changes are 3–4 °C [20, 21]. Wind speed for the heating period averaged 3.6 m/s with predominantly westerly and southerly and southeasterly wind directions.

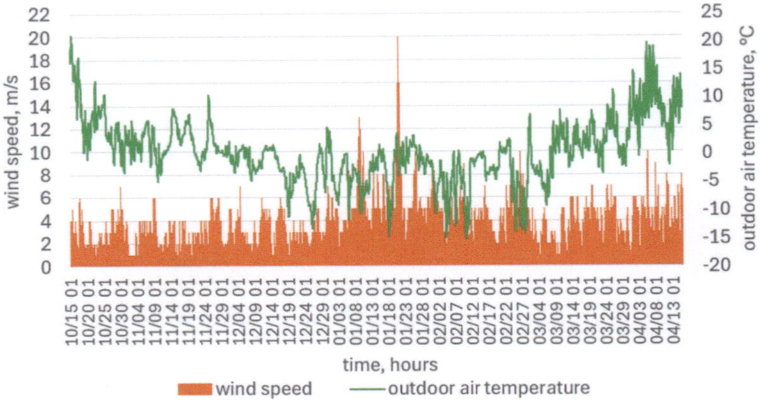

Fig. 4.36 Hourly temperature and wind speed during the heating period

Hourly solar heat gain on vertical surfaces for the southern orientation is shown in Fig. 4.37. The radiation transmittance of the windows is 0.691, which was calculated using the EnergyPlus software product [17].

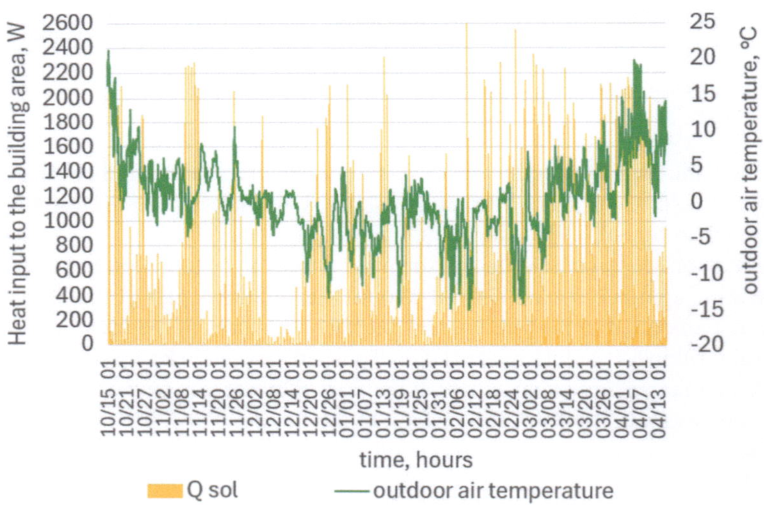

Fig. 4.37 Hourly solar heat gain in a representative apartment during the heating period

4.6 Energy Consumption of a Building with Regard to Natural Air Exchange According to Different Standards

As part of the study, the impact of the air exchange mode on the heat load and energy consumption of the facility was investigated. For this purpose, we used a dynamic model of an apartment in the EnergyPlus software environment [17], which allowed us to conduct hourly simulations. The energy model of the apartment was developed taking into account the operating mode and internal heat gains, including people, lighting and electrical equipment. This allowed for more accurate and realistic results.

The results of the calculations were presented for the climatic conditions of a typical year, which were taken from the international climatological database IWEC [20] for the city of Kyiv. This made it possible to take into account seasonal variations and the specific climate of this geographical area.

Energy consumption for residential buildings in Ukraine is calculated on the basis of the standards DBN B.2.2-15:2019 "Residential buildings. Basic Provisions" [22] and DSTU 9091:2022 "Energy Efficiency of Buildings" [19].

According to the standard DBN B.2.2-15:2019 "Residential buildings. Basic Provisions" [22], which is used for design, reconstruction and overhaul conditions, the minimum air exchange rate in houses with natural ventilation for common rooms, bedrooms, children's rooms and offices is 0.5 h^{-1}, for kitchens this value is 1.5 h^{-1}. In the process of energy certification of buildings, energy auditors assume that the air exchange rate in residential buildings with natural ventilation is 0.6 h^{-1} in accordance with the standard [5]. However, in the calculations of the energy certificate of buildings, this value is taken as infiltration.

To date, most existing standards do not have a clearly defined method for establishing the minimum air exchange rate, but only offer recommendations on how to determine this parameter. In accordance with ASHRAE 62.2-2022, the air exchange rate for the entire apartment is determined by the formula (4.4):

$$Q_{tot} = 0.15A_{floor} + 3.5(N_{br} + 1), \tag{4.4}$$

where Q_{tot}—total required level of air exchange, l/s; A_{floor}—housing area, m^2; N_{br}—number of bedrooms.

The air volume calculated by formula (4.4) for a one-bedroom apartment with a total area of 32.8 m^2 is 11.92 L per second. This corresponds to an air exchange rate of 0.52 h^{-1}.

According to the EN 15,251:2011 standard, the assessment of the indoor environment of the apartment under study belongs to category II, which corresponds to the normal level of expectations. To calculate the air exchange rate in a representative apartment, an air flow rate of 0.42 l/s per square metre is required [23], which is equivalent to an air exchange rate of 0.6 h^{-1} at an internal height of 2.5 m, which is the same as the table values of this standard.

The results of energy consumption for heating needs for an apartment under conditions of normative air exchange in accordance with national and international standards are shown in annual (Fig. 4.38) and hourly (Fig. 4.39) sections.

For the modelling variant in accordance with ASHRAE 62.2-2022, the calculated value of the air exchange rate was taken at 0.52 h^{-1}.

The results demonstrate that the annual heat consumption of a representative apartment can vary significantly depending on the requirements for air exchange rates set out in different standards.

In particular, it was found that when using the DBN B.2.2-15:2019 standard and DSTU B EN 15,251:2011, the difference in annual heat consumption is 2.4%.

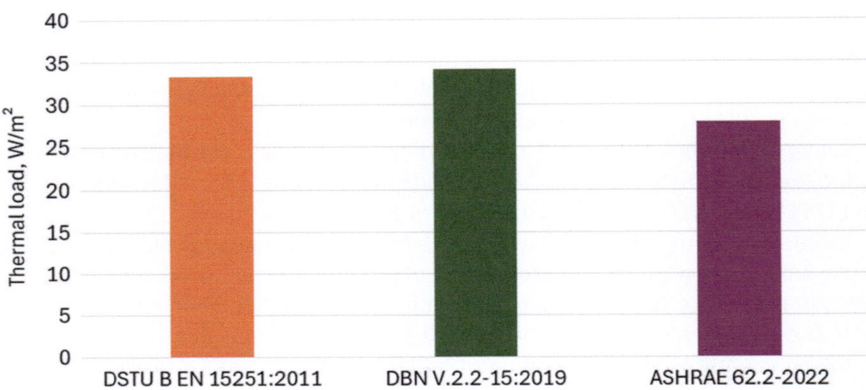

Fig. 4.38 Annual heat load of an apartment

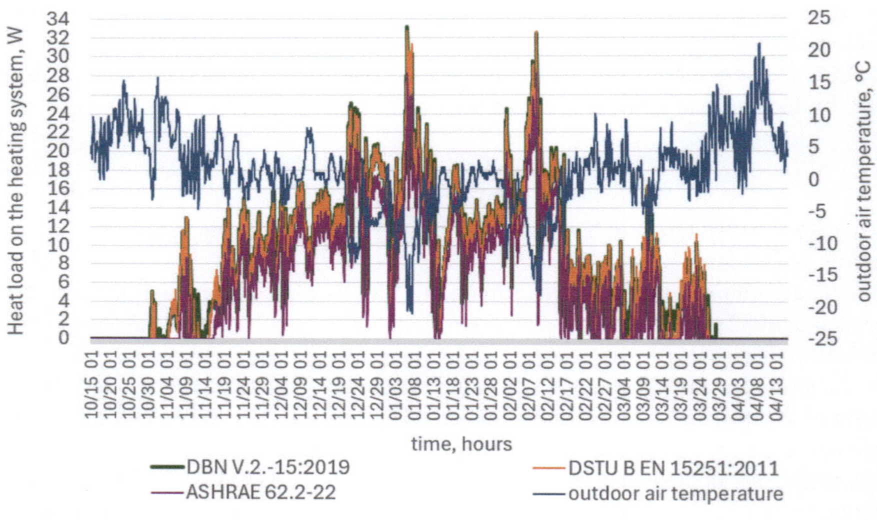

Fig. 4.39 Hourly heat consumption of an apartment

In addition, when comparing the standards DBN B.2.2-15:2019 and ASHRAE 62.2-2022, the difference in annual heat consumption reaches 18%. This indicates a significant difference in the requirements for air exchange rates set out in these standards and their impact on energy consumption.

Thus, the choice of the standard whose air exchange rate requirements will be used in the calculations can have a significant impact on the energy efficiency of the facility.

4.7 Energy Consumption of the Building with Regard to Changes in Natural Air Exchange

Based on the experiment, the main mechanisms of air movement in the premises of a typical apartment were identified, depending on the variability of external conditions.

Taking into account the mechanisms of air movement, the following dependence is established. If fresh air enters the living room and kitchen from the outside (air movement mechanism I), it is heated and when it is moved to other rooms of the apartment, the energy for heating this air does not need to be spent again. When fresh air is supplied through the ventilation duct (air movement mechanism II), it also does not need to be heated, as it is heated directly in the ventilation duct. Therefore, the study took into account the dependence of the change in the mechanism of air transfer between the rooms of the apartment depending on the wind direction.

The climate file for a typical IWEC year contains hourly wind speed and direction data (Fig. 4.40). The wind direction is given in degrees. Figure 4.41 shows the feature of determining the wind direction according to the angle.

According to the experimental data, in the northeast, east, southeast and southwest wind directions, the infiltration phenomenon (air movement mechanism I) prevails in

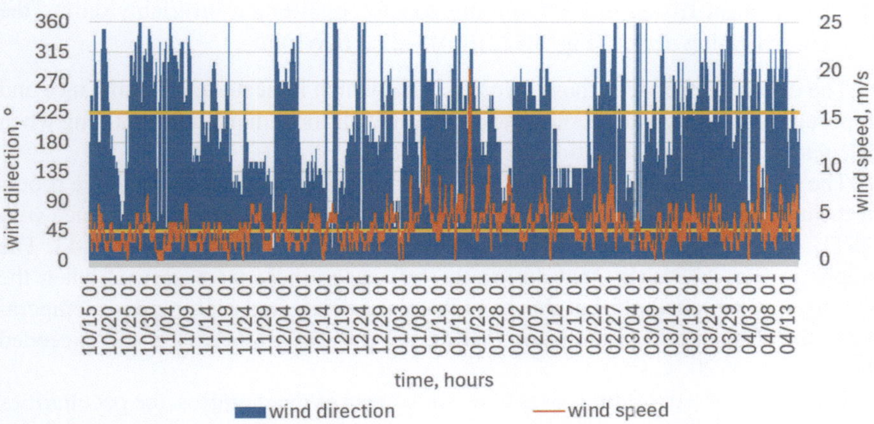

Fig. 4.40 Hourly variation of wind speed and direction: I—infiltration, II—exfiltration

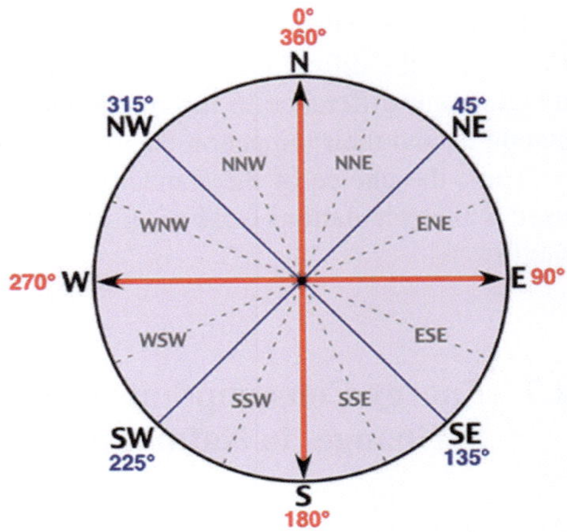

Fig. 4.41 Determination of the wind direction given by the angle according to the IWEC typical year file [20]

the premises of an apartment with external walls oriented to the south. Conversely, in the case of western, north-western, northern wind directions or calm, the phenomenon of exfiltration (air movement mechanism II) was predominantly observed. In the energy modelling, according to the experimental data, the air exchange rate was set as an average value with hourly consideration of the wind direction.

The current standards of Ukraine [5, 22, 23] do not take into account the direction of air movement. Therefore, we calculated the experimentally established levels of natural air exchange, provided that the direction/mechanism of supply air movement in the apartment is not taken into account.

The following options were considered for comparing the marginal cases of energy consumption for heating needs:

(1) Mechanism I—air enters the room zone by infiltration invariably during the entire heating period (Fig. 4.42, I(a))—78.8 (kWh)/m^2;
(2) Mechanism II—air enters the room zone by exfiltration invariably during the entire heating period (Fig. 4.42, II(a))—23.3 (kWh)/m^2.

The difference between these values is 70%, which indicates the significance and importance of taking into account the direction of air movement in the building when calculating energy consumption for heating needs.

The dynamics of internal and radiation temperatures in the representative rooms of the apartment using a constant value of the air exchange rate in accordance with DSTU B EN 15,251:2011 and the actual air exchange is shown in Fig. 4.43. The graphs in Fig. 4.43 show that during the off-season there were periods when the heating was switched off, but due to high solar activity and high outdoor temperatures, the temperature in the apartment during this period increased and exceeded the standard of 20 °C [19].

In the considered limiting cases of air movement in the premises, the peculiarities of supply air heating are also taken into account. That is, the case when the outside air entering the apartment is heated, and during the subsequent mass transfer of air

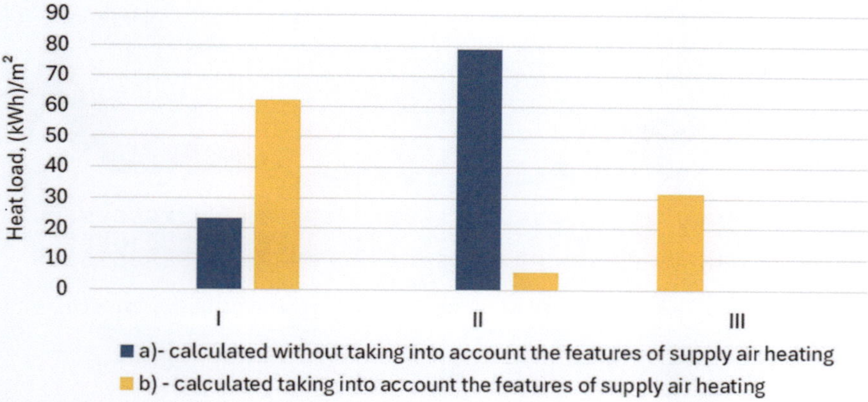

Fig. 4.42 Annual heat load on the heating system: I—the infiltration mechanism is constant throughout the heating period; II—the exfiltration mechanism is constant throughout the heating period; III—taking into account the hourly change in natural air exchange

to other zones (rooms) of the apartment, energy is not consumed for its heating (Fig. 4.42, data series (b)). And the case when the energy for air heating is consumed in each zone of the apartment (Fig. 4.42, data series (a)).

In the two extreme cases, when taking into account the peculiarities of supply air heating, the annual heat consumption of a representative apartment for infiltration conditions during the entire heating period will decrease by 21%, and for exfiltration conditions during the entire heating period will decrease by 76%.

Figure 4.42III also shows the heat load of an apartment, taking into account the variability of air movement mechanisms over time, depending on weather conditions. This value is close to the average value between the extreme cases and is 31.6 (kWh)/ m^2.

Figure 4.44 shows the results of modelling energy consumption for heating needs for a representative apartment, taking into account the level of natural air exchange rate in accordance with regulatory documents and experimental data.

According to the latest published data from the State Statistics Service of Ukraine, the average final energy consumption by households in Ukraine in 2016–2020 is 15576.8 ktoe. At the same time, household energy consumption for heating purposes averaged 44% in 2016–2020, which is equivalent to 6814.2 ktoe.

According to the results of energy modelling, which was carried out taking into account climate data and the dependencies determined during the experiments, the annual energy savings compared to the standard used for energy certification of buildings DSTU B EN 15,251:2011 is 5.4%. In recalculation according to the statistical data for 2016–2020, the annual heat energy savings are 3.7 million Gcal. Compared to the standard used for the design, reconstruction and overhaul of buildings, DBN B.2.2-15:2019, the savings will be 7.7%, which is equivalent to 5.2 million Gcal. Gcal.

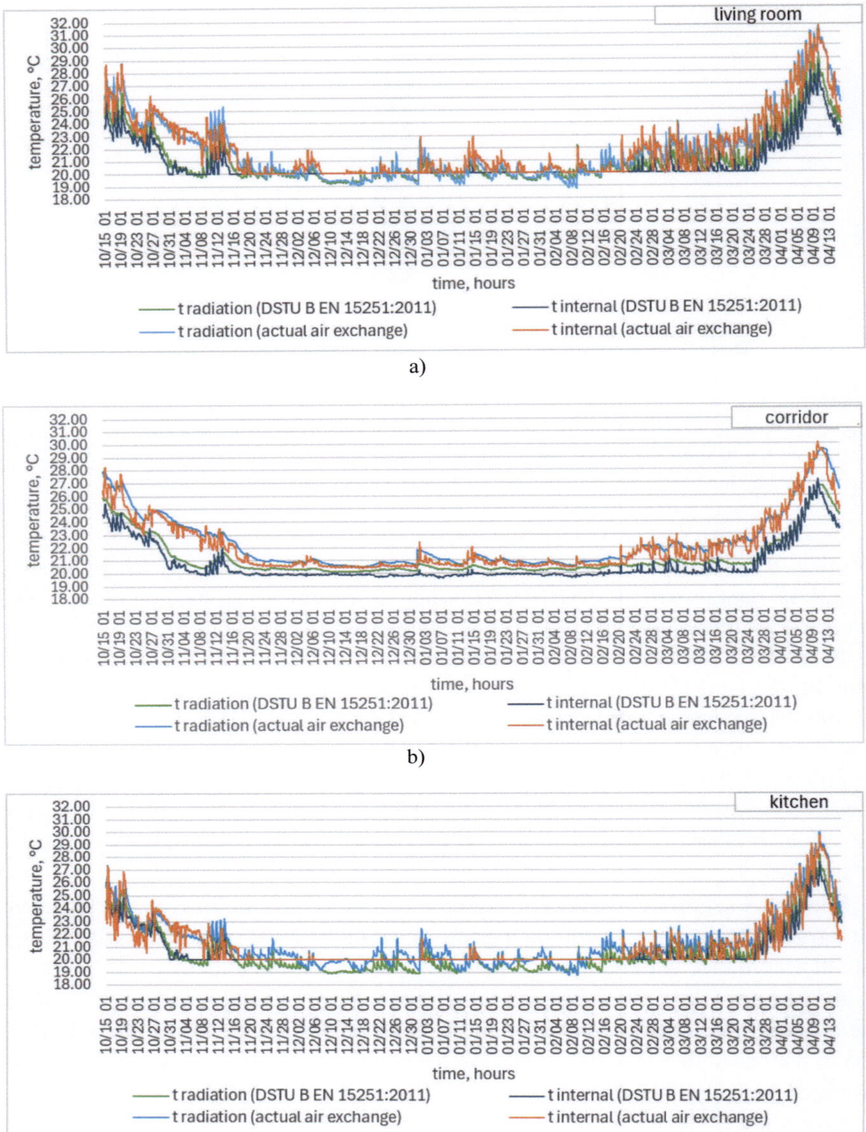

Fig. 4.43 Hourly changes in internal and radiation temperatures in the living room (**a**), apartment corridor (**b**) and kitchen (**c**)

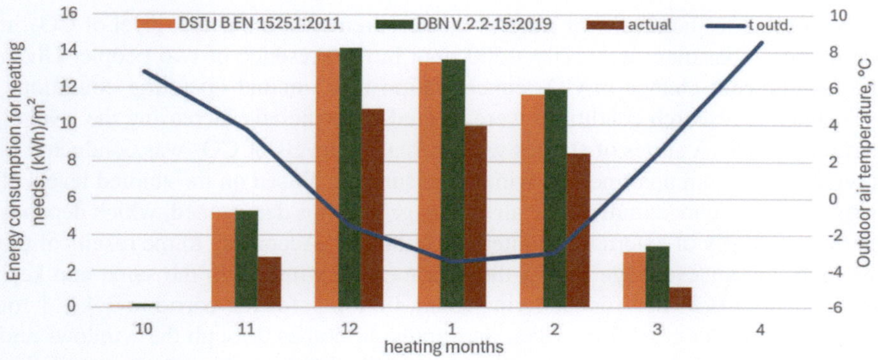

Fig. 4.44 Monthly energy consumption for heating needs

This part of the study focused on the influence of the air exchange mode on the heat load and energy consumption of the object. For this purpose, a 3D building model of the studied one-room apartment was developed in the DesignBuilder software [18], and a dynamic model of the apartment was created in the EnergyPlus software environment [17], which was used for hourly simulation. In the process of creating the energy model, the apartment space was divided into five zones: living room, corridor, kitchen, bathroom, and storage room. The model takes into account the operating mode of the apartment and internal heat sources such as people, lighting and electrical equipment.

As a result, the energy consumption for heating an apartment was determined, provided that the normative air exchange established by national and international standards is observed, as well as taking into account the actual air exchange rate determined experimentally. According to the results of energy modelling,, the annual energy savings, taking into account the actual values of the natural air exchange rate in different areas of the building, is 5.4% compared to the standard DSTU B EN 15,251:2011, and 7.7% compared to the standard DBN B.2.2-15:2019.

4.8 Conclusions

According to the results of the experimental determination of the CO_2 concentration, it was found that the average air exchange rate in the classrooms of an educational institution in winter is 0.41 h^{-1}, taking into account airing, the average air exchange rate is 0.48 h^{-1}.

The results of the study of CO concentration$_2$ in the classroom of the educational building №22 of the Igor Sikorsky Kyiv Polytechnic Institute in the cooling mode indicate the dependence of the value of the natural air exchange rate on the wind and stack effects. As the temperature difference between indoor and outdoor air increased, the air exchange rate increased from 0.37 to 0.9 h^{-1}.

According to the results of a series of measurements, the actual level of CO_2 in a one-room apartment was 1200–1500 ppm in the presence of two people. Given the insignificant change in CO_2 concentration under actual operating conditions, experimental research conditions were created by artificially increasing the level of carbon dioxide. A series of studies with increasing levels of CO_2 was conducted in a typical one-room apartment in winter and summer. Based on the studied levels of carbon dioxide concentration, the air exchange rate was determined, which depends on the variability of external and internal conditions. According to the results of the research, the air exchange rate in the apartment in winter was the same and kept at the level of 0.74 h^{-1} for the living room, 0.73 h^{-1} for the corridor, 3.4 h^{-1} for the kitchen, under conditions when the supply air comes through the windows and the front door. For the conditions when the supply air comes through the ventilation duct and the front door, the air exchange rate for the living room was 0.37 h^{-1}, for the corridor—1.06 h^{-1}, for the kitchen—0.57 h^{-1}. For the summer period, the determined air exchange rate for the infiltration mechanism was 0.57 h^{-1} for the living room, 2.34 h^{-1} for the corridor, 2.41 h^{-1} for the kitchen. With exfiltration air movement, the average air exchange rate for the living room was 0.24, 0.94 h^{-1} for the corridor, 0.52 h^{-1} for the kitchen.

The analysis of quantitative data on the air exchange rate obtained during observations in summer and winter indicates the importance of taking into account seasonal variations. It was found that in summer, natural air exchange mainly occurs under the influence of the wind effect. During the heating season, air exchange is mainly driven by a combination of the stack and wind effects.

As a result of dynamic modelling in the DesignBuilder and EnergyPlus software, the energy consumption for heating the apartment was determined, subject to compliance with the normative air exchange established by national and international standards, as well as taking into account the actual air exchange rate determined experimentally. According to the results of energy modelling, the annual energy savings, taking into account the actual values of the natural air exchange rate in different areas of the building, are 5.4% compared to the DSTU B EN 15,251:2011 standard, and 7.7% compared to the DBN B.2.2-15:2019 standard.

References

1. Deshko, V., Bilous, I., Sukhodub, I., Yatsenko, O.: Evaluation of energy use for heating in residential building under the influence of air exchange modes. J. Build. Eng. **42**, 103020 (2021). https://doi.org/10.1016/j.jobe.2021.103020
2. American society of heating refrigerating and air-conditioning engineers. In: 2013 ASHRAE Handbook: Fundamentals. ASHRAE, Atlanta, GA (2013)
3. Standard guide for using indoor carbon dioxide concentrations to evaluate indoor air quality and ventilation. Replaces ASTM D6245-12; effective from 2023-06-01, Official edn. American Society of Heating, Refrigerating and Air-Conditioning Engineers, West Conshohocken, PA (2018)

4. Pereira, P.F., Ramos, N.M., Almeida, R.M., Simoes, M.L., Barreira, E.: Occupant influence on residential ventilation patterns in mild climate conditions. Energy Procedia **132**, 837–842 (2017)
5. Heating, ventilation and air conditioning, DBN V.2.5-67:2013, Ministry of Regional Development of Ukraine, Kyiv (2013)
6. Design parameters of indoor microclimate for the design and assessment of energy performance of buildings in relation to air quality, thermal comfort, lighting and acoustics (EN 15251:2007, IDT), DSTU B EN 15251:2011. State Enterprise "Ukrarchbudinform", Kyiv (2012)
7. Deshko, V., Bilous, I., Vynogradov-Saltykov, V., Shovkaliuk, M., Hetmanchuk, H.: Integrated approaches to determination of co2 concentration and air rate exchange in educational institution. Rocznik Ochrona Srodowiska **22**(1), 82–104 (2020). https://ros.edu.pl/images/roczniki/2020/007_ROS_V22_R2020.pdf
8. Lu, S., Zheng, S., Kong, X.: The performance and analysis of office building energy consumption in the west of Inner Mongolia Autonomous Region, China. Energy Build. **127**, 499–511 (2016)
9. Bilous, I., Biriukov, D., Karpenko, D., Eutukhova, T., Novoseltsev, O., Voloshchuk, V.: Reinforcement learning model for energy system management to ensure energy efficiency and comfort in buildings. Energy Eng. https://doi.org/10.32604/ee.2024.051684
10. Sinoptik. https://ua.sinoptik.ua/. Accessed 23 Dec 2023
11. Bilous, I.Y., Deshko, V.I., Sukhodub, I.O.: Building energy modelling using hourly infiltration rate. J. Civ. Eng. **96**(4), 27–41 (2020)
12. Bilous, I., Biriukov, D., Eutukhova, T., Novoseltsev, O., Voloshchuk, V., Prozuments, A.: Energy efficiency and indoor climatic conditions in buildings: dynamic modeling for systemic improvement. Sec. Indoor Environ. **10** (2024). | https://doi.org/10.3389/fbuil.2024.1502794
13. Deshko, V., Bilous, I., Buyak, N., Shevchenko, O.: The impact of energy-efficient heating modes on human body exergy consumption in public buildings. In: 2020 IEEE 7th International Conference on Energy Smart Systems, ESS 2020 - Proceedings, c. 201–205, 9160270 (2020). https://www.scopus.com/record/display.uri?eid=2-s2.0-85092164581&origin=resultslist
14. Buyak, N., Deshko, V., Bilous, I., Biriukov, D., Voloshchuk, V.: Applying dynamic energy and exergy analysis to a building envelope. Int. J. Exergy **43**(2), 93–107 (2024). https://doi.org/10.1504/IJEX.2024.136928
15. Deshko, V., Bilous, I., Boiko, T., Shevchenko, O., Borodinecs, A., Zemitis, J.: Energy performance of higher education institutions buildings operating during quarantine restrictions and/or martial law in Ukraine. Latv. J. Phys. Techn. Sci. **2**, 44–65 (2024). https://doi.org/10.2478/lpts-2024-0013
16. Deshko, V.I., Bilous, I.Y., Sukhodub, I.O.: Building energy modelling using hourly infiltration rate. J. Civ. Eng. **6**(4), 27–41 (2020)
17. "EnergyPlus. Date of access: 7 December. 2024. [Online]. Available: https://energyplus.net/.
18. DesignBuilder. https://designbuilder.co.uk/. Accessed 7 Dec 2024
19. Energy efficiency of buildings. Method for calculating energy consumption during heating, cooling, ventilation, lighting and hot water supply, DSTU 9190:2022, SE NDIBK, Kyiv (2022)
20. International Weather for Energy Calculations. https://energyplus.net/weather-location/europe_wmo_region_6/UKR
21. DSTU-N B V.1.1-27: Construction climatology. Ministry of Regional Development of Ukraine, Kyiv (2010)
22. Residential buildings. Basic provisions, DBN B.2.2-15:2019, Ministry of Regional Development, Construction, Housing and Communal Services of Ukraine, Kyiv (2019)
23. Estimated parameters of indoor microclimate for the design and assessment of energy performance of buildings in relation to air quality, thermal comfort, lighting and acoustics (EN 15251:2007, IDT), DSTU B EN 15251:2011. State Enterprise "Ukrarkhbudinform", Kyiv (2012)

Appendix A
Sensor Nomenclature and Results of Experimental Studies of Their Characteristics

See Figs. A.1, A.2, A.3, A.4 and Tables A.1, A.2.

A. Zaporozhets et al., *Buildings Energy Audit*, Studies in Systems, Decision and Control 641, https://doi.org/10.1007/978-3-032-07200-9

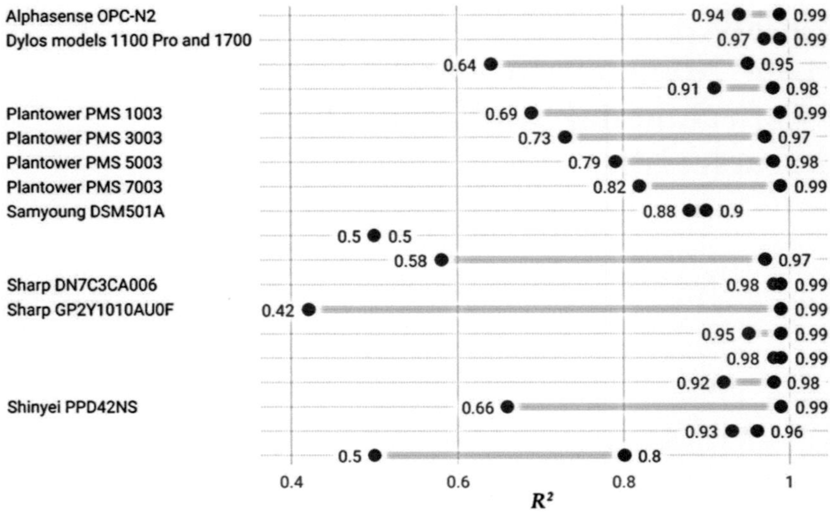

a)

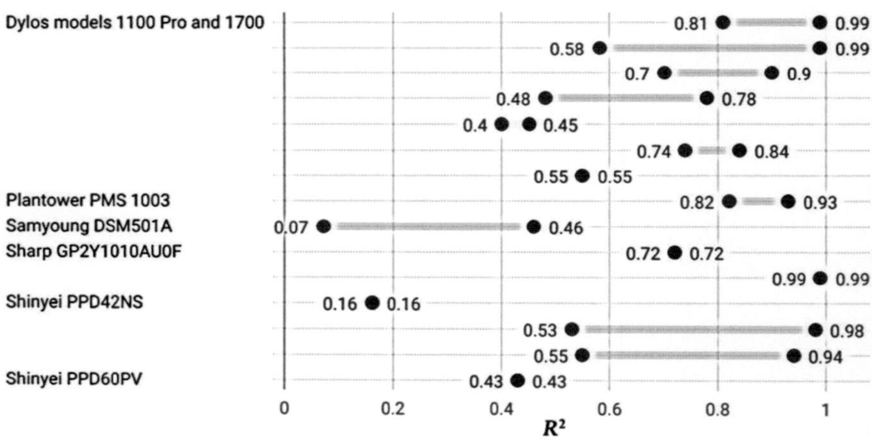

b)

Fig. A.1 Comparison of PM sensor measurement data with the data of reference devices in laboratory (**a**) and field (**b**) conditions based on R^2

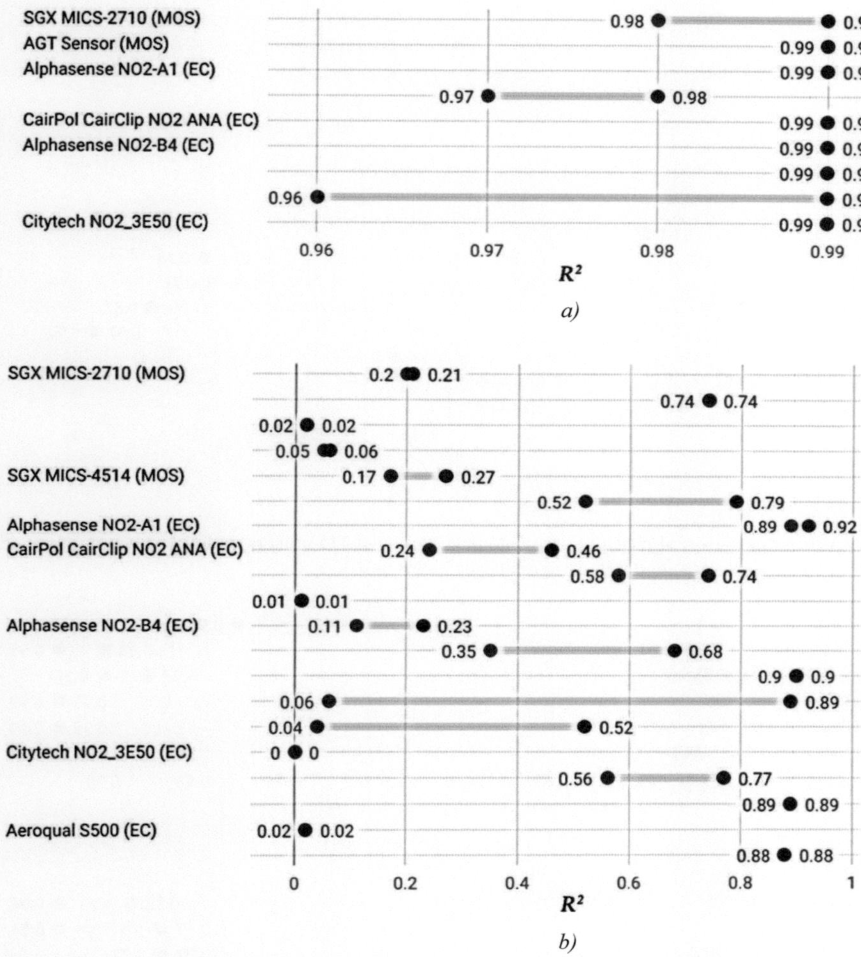

Fig. A.2 Comparison of measurement data of low-cost NO sensors$_2$ with the data of the reference device in laboratory (**a**) and field (**b**) conditions based on R^2

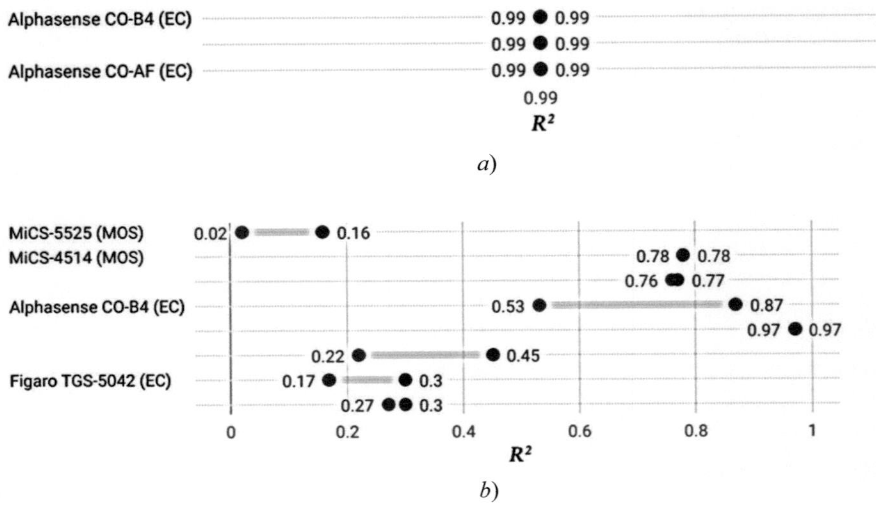

Fig. A.3 Comparison of measurement data from low-cost CO sensors with data from *a* reference device in laboratory (**a**) and field (**b**) conditions based on R^2

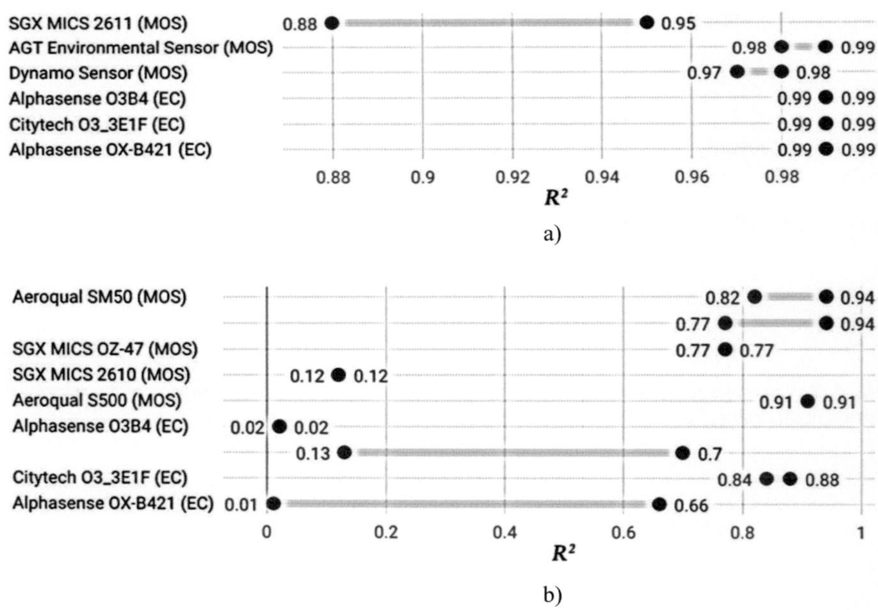

Fig. A.4 Comparison of the measurement data of low-cost sensors O_3 with the data of the reference device in laboratory (**a**) and field (**b**) conditions based on R^2

Table A.1 Models of sensors and devices that can be used for indoor air pollution monitoring and their selected characteristics

№	Pollutant	Model	Method	Resolution	Error	Measuring range
1	$PM_{2.5}$	Met one instrument BAM-1020	Beta attenuation	1 $\mu g/m^3$	± 1 $\mu g/m^3$	0–1000 $\mu g/m^3$
2		Met one instrument aerocet 831	Light scattering	0.1 $\mu g/m^3$	$\pm 10\%$	0–1000 $\mu g/m^3$
3		Alphasense OPC-N2	Light scattering	–	–	–
4		Sharp microelectronics DN7C3CA006	Light eclipse (nephelometry)	–	–	25–500 $\mu g/m^3$
5	PM_{10}	Teledyne model 602 BetaPLUS	Beta attenuation	0.1 $\mu g/m^3$	± 1 $\mu g/m^3$	0–1500 $\mu g/m^3$
6		Met one instrument aerocet 831	Light scattering	0.1 $\mu g/m^3$	$\pm 10\%$	0–1000 $\mu g/m^3$
7		Alphasense OPC-N2	Light scattering	–	–	0.1–1,500,000 $\mu g/m^3$
8		Sharp GP2Y1010AU	Light eclipse (nephelometry)	–	–	0–600 $\mu g/m^3$
9	NO_2	Teledyne model T500U	Resonant attenuated phase shift spectroscopy	0.1 ppb	$\pm 0.5\%$	0–5 ppb 0–1 ppm
10		Aeroqual series 500	Electrochemical	1 ppb	± 0.02 ppm at 0–0.2 ppm or $\pm 10\%$ at 0.2–1 ppm	0–1 ppm
11		Alphasense B4 series	Electrochemical	12 ppb	–	0–20 ppm
12		SGXSensorTech MiCS-2714	Solid state sensor	–	–	0.05–10 ppm
13	SO_2	Teledyne model 6400 T/6400E	UV fluorescence	0.1 ppb	$\pm 0.5\%$	0–50 ppb or 0–20 ppm
14		Aeroqual series 500	Electrochemical	10 ppb	± 0.05 ppm at 0–0.5 ppm or $\pm 10\%$ at 0.5–10 ppm	0–10 ppm
15		Alphasense B4 series	Electrochemical	5 ppb	–	0–100 ppm
16		Hanwei MQ-136	Solid state sensor	–	–	0–200 ppm

(continued)

Table A.1 (continued)

№	Pollutant	Model	Method	Resolution	Error	Measuring range
17	CO	Teledyne model T300U	Infrared	0.1 ppb	±0.05%	0–100 ppb or 0–100 ppm
18		Aeroqual series 500	Electrochemical	10 ppb	±0.5 ppm at 0–5 ppm or ±10% at 5–25 ppm	0–25 ppm
19		Alphasense B4 series	Electrochemical	4 ppb	–	0–1000 ppm
20		Hanwei MQ-7	Solid state sensor	–	–	20–2000 ppm
21		e2V MiCS-5525	Solid state sensor	–	–	0–1000 ppm
22		Hanwei MQ-9	Solid state sensor	–	–	0–1000 ppm
23	O_3	Teledyne model 265E	Chemiluminescent	0.1 ppb	±0.5%	0–100 ppb or 0–2 ppm
24		Aeroqual Series 500	Solid state sensor	1 ppb	±5 ppb	0–150 ppb
25		Alphasense B4 series	Electrochemical	4 ppb	–	0–5 ppm
26		Hanwei MQ-131	Solid state sensor	–	–	10–1000 ppm
27	CO_2	Winson MG811	Solid state sensor	–	–	350–10,000 ppm
28		GE/ 6004/6113	Infrared	–	±40 ppm	0–2000 ppm
29		ELT H550	Infrared	–	±5%	0–10,000 ppm
30		FIGARO CDM4160	Solid state sensor	–	±20%	400–45,000 ppm
31		GSS C20/C100	Infrared	–	±10 ppm	0–5000 ppm
32	ENT	Alphasense AH2	Photoionisation	–	±3%	0–40 ppm
33		Alphasense A12	Photoionisation	–	±5%	0–4000 ppm

Table A.2 PM sensor models and their selected characteristics

№	Model	Dimensions, mm	Weight, g	Power supply	Max. current consumption, mA	Cost, $, per person	Particle size	Measuring range
1	Alphasense OPC-N2	75 × 64 × 60	105	5 V DC	175	~500	0.38–17 μm in 16 ranges	0.1–1,500,000 μg/m^3
2	Dylos DC 1100 Pro	178 × 114 × 76	544	110 V AC	–	~300	0.5–2.5 μm; 0.5–10 μm	0–106 p/cm^3
3	Dylos DC 1700	178 × 114 × 76	544	110 V AC	–	~400	0.5–2.5 μm; 0.5–10 μm	0–106 p/cm^3
4	Novafitness SDL301	204 × 100 × 36	580	5 V DC	–	~250	0.3–2.5 μm; 0.3–10 μm	0–1000 μg/m^3
5	Novafitness SDL607	73 × 73 × 20	120	5 V DC	–	~120	0.3–2.5 μm; 0.3–10 μm	0–1000 μg/m^3
6	Novafitness SDS011	71 × 70 × 23	–	5 V DC	80	~35	0.3–2.5 μm; 0.3–10 μm	0–1000 μg/m^3
7	Novafitness SDS018	59 × 45 × 20	–	5 V DC	70	~40	0.3–2.5 μm; 0.3–10 μm	0–1000 μg/m^3
8	Novafitness SDS021	42 × 32 × 24	–	5 V DC	70	~35	0.3–2.5 μm; 0.3–10 μm	0–1000 μg/m^3
9	Novafitness SDS198	71 × 70 × 23	–	5 V DC	80	~80	1–100 μm	0–20,000 μg/m^3
10	Plantower PMS 1003	65 × 42 × 23	–	5 V DC	120	~20	0.3–1 μm; 1–2.5 μm; 2.5–10 μm	0–500 μg/m^3

Appendix B
Thermal Performance of Building Envelopes and Heating Systems

See Tables B.1, B.2, B.3 and B.4.

Table B.1 Minimum permissible value of the reduced heat transfer resistance of the building envelope of residential and public buildings R_{qmin}

No.	Type of enclosing structure	$R_{q\min}$, $m^2 \cdot K/W$, for the temperature zone	
		I	II
1	External wall enclosing structures	4.00	3.50
2	Combined coatings adjacent to the outside by air	7.00	6.00
3	Coating of heated attics (technical floors), attics, attic floors of unheated attics	6.00	5.50
4	Ceilings bordering on the outside air and over unheated basements	5.00	4.00
5	Translucent enclosing structures	0.90	0.70
6	Anti-aircraft lights	0.80	0.70
7	External doors	0.70	0.60

A. Zaporozhets et al., *Buildings Energy Audit*, Studies in Systems, Decision and Control 641, https://doi.org/10.1007/978-3-032-07200-9

Table B.2 National values for internal heat capacity

Class	Internal heat capacity of the building, W·h/(m²·K)	Outside air temperature, °C, which is taken as the design temperature
Very lightweight	25	$t_{з.п.min}$
Lightweight	35	$t_{з.п.1}{}^{0.98}$
Medium	50	$t_{з.п.1}{}^{0.92}$
Heavy	80	$\frac{t_{з.п.1}{}^{0.92} + t_{з.п.5}}{2}$
Very heavy	110	$t_{з.п.5}$

Table B.3 Efficiency of free-flowing heating surfaces (radiators); rooms with a height of no more than 4 m

Influencing factor		Components of the overall level of efficiency			
		η_{str1}	η_{str2}	η_{ctr}	η_{emb}
Controlling the room air temperature	None			0.86	
	At the average (typical) air temperature of the building premises			0.88	
	P-regulation* (2 K)**			0.93	
	P-regulation* (1 K)**			0.95	
	PI control***			0.97	
	PI control*** with optimisation (e.g. dispatching, adapted control)			0.99	
Temperature head (at an air temperature of 20° C)	60 K (e.g. 90/70)	0.88			
	42.5 K (e.g. 70/55)	0.93			
	30 K (e.g. 55/45)	0.95			
Specific heat loss through external enclosures	The heater is installed against an internal wall		0.87		1
	The heater is installed against an external wall:				
	– a window without radiation protection		0.83		1
	– a window with radiation protection		0.88		1
	– ordinary wall		0.95		1

* Proportional regulation
** Temperature control accuracy (in Kelvin degrees)
*** Proportional integral control

Table B.4 Influence of the method of balancing the heating system on heat energy savings

System	Influencing factors	Accounting factor hydraulic balancing systems f_{hydr}	Saving energy
Single-pipe (permanent hydraulic mode)	The system is not adjusted. There are no balancing valves on the risers (horizontal branches) of the system	1.09	Basic version comparison of systems 0
	The system is set up. There are manual balancing valves on the risers (horizontal branches)	1.07	8.3%
	The system is well established. There are automatic flow controllers (stabilisers) on the risers (horizontal branches)	1.05	16.6%
Single-pipe (variable hydraulic mode)	The system is set up. There are automatic flow controllers (limiters) with stabilisation of the coolant temperature at the outlet of the riser (horizontal branch)	1.01	33.1%
	The system has been adjusted. There are automatic flow controllers (limiters) with temperature control of the heat carrier at the outlet of the riser (horizontal branch) according to the temperature schedule	1.00	8.3%

MIX
Papier aus verantwortungsvollen Quellen
Paper from responsible sources
FSC® C105338

If you have any concerns about our products,
you can contact us on
ProductSafety@springernature.com

In case Publisher is established outside the EU,
the EU authorized representative is:
Springer Nature Customer Service Center GmbH
Europaplatz 3, 69115 Heidelberg, Germany

Printed by Libri Plureos GmbH
in Hamburg, Germany